Advanced Structured Materials

Volume 37

Series editors

Andreas Oechsner, Southport, Australia
Lucas F. M. da Silva, Porto, Portugal
Holm Altenbach, Magdeburg, Germany

For further volumes:
http://www.springer.com/series/8611

Mohd N. Tamin · Norhashimah M. Shaffiar

Solder Joint Reliability Assessment

Finite Element Simulation Methodology

Springer

Mohd N. Tamin
Faculty of Mechanical Engineering
Universiti Teknologi Malaysia
Skudai, Johor
Malaysia

Norhashimah M. Shaffiar
Department of Manufacturing
and Materials
Kulliyyah of Engineering
International Islamic University Malaysia
Gombak, Kuala Lumpur
Malaysia

ISSN 1869-8433 ISSN 1869-8441 (electronic)
ISBN 978-3-319-00091-6 ISBN 978-3-319-00092-3 (eBook)
DOI 10.1007/978-3-319-00092-3
Springer Cham Heidelberg New York Dordrecht London

Library of Congress Control Number: 2014936614

Printed on acid-free paper

Springer is part of Springer Science+Business Media (www.springer.com)

Preface

This book addresses various aspects of solder joints' reliability assessment for microelectronic assembly with Ball Grid Array (BGA) solder joints. Computational aspects of the reliability assessment are deliberated using Finite Element (FE) simulation methodology. The discussion begins with a systematic approach in engineering problem solving with the aid of FE simulation. The approach is elaborated based on numerical experiment framework. Essential requirements for FE simulation of engineering phenomena are described. These cover assumptions considered for a simplified physical model, FE model geometry development, constitutive models for materials used, and aspects of FE model validation and interpretation of numerical results. The explanation is further illustrated using an example of procedures for FE simulation of microelectronic assembly with BGA solder joints subjected to reliability test conditions. Fundamentals of the mechanics of solder material are adequately reviewed in relation to computational aspects of FE formulations. Constitutive models for solder alloys representing elastic-plastic response, creep deformation, and low cycle fatigue behavior are adequately described. Concept of damage is introduced along with deliberation of continuum and local approach in damage mechanics modeling. Cohesive zone model and continuum damage model is described for failure simulation of solder/intermetallic interface and bulk solder joint material, respectively. Extension of these models to account for cyclic loading cases is deliberated. Hybrid experimental–computational approach for determination of damage model parameters is presented. Applications of the deliberated methodology to selected problems in assessing reliability of solder joints in microelectronic assemblies are demonstrated. These industry-defined research-based problems related to solder joint reliability assessment include solder reflow cooling, temperature cycling, and mechanical fatigue of a test assembly with BGA solder joints, JEDEC board-level drop test, and shear fatigue of reflowed solder specimen. Emphasis is placed on accurate quantitative assessment of solder joint reliability through basic understanding of the mechanics of materials as interpreted from results of FE simulations. Validation aspects of the FE simulation including FE models and material constitutive models are equally considered. The FE simulation methodology, although demonstrated for reliability assessment of solder joints, in this book is readily applicable to numerous other problems in mechanics of materials and structures.

The book is a compilation of research experience and training in the development of validated methodology for solder joint reliability assessment based on the mechanics of reflowed solder system, typically found in Plastic Ball Grid Array (PBGA) packages and assemblies. FE models of microelectronic test assembly employed in various examples throughout this book are based on typical geometry and grid array of the solder joints. However, the simulation methodology is equally valid for different geometry and design of solder array. In addition, the FE simulation methodology deliberated in this book is software-independent although all FE results are presented using graphical output of Abaqus FEA software employed in running the FE simulation cases. The coverage and arrangement of the topics are intended to serve as a reference book or course text on Computational Methodology for Solder Joint Reliability at the undergraduate senior or graduate level. Sufficient review of essential fundamental elements of the mechanics of materials and practical coverage of damage mechanics renders the book an excellent reference for practicing engineers, technical personnel, academicians, and researchers involved in FE simulations (problem identification, modeling, and analysis) and assessment of solder joint reliability in microelectronic packages and assemblies.

Layout of the Book

The chapters in this book are arranged to address the two aspects of the contents, namely FE simulation methodology and illustration of the methodology in the assessment of solder joint rcliability for microelectronic assemblies with BGA solder joints.

Chapter 1 begins with an overview of the trend in electronic packaging technology. It also describes an electronic assembly with a flip chip package and ball grid array solder joints. The terms *reliability assessment* and *reliability simulation* of the assembly are briefly described.

Chapter 2 provides an overview of the simulation methodology as applied in engineering problem solving. The various aspects of engineering simulation including the real-life problem of interest, available numerical tools, relevant physical laws, and experimental data generation are described within a numerical experiment framework. A short review on derivation of finite element equations is included. Typical process flow in FE-based solution approach to a well-identified problem is discussed.

Chapter 3 outlines and discusses the essential requirements for solving engineering problems through FE simulation. These include FE modeling processes (model geometry, element discretization, boundary conditions, and loading), accuracy of constitutive models, and aspects of FE model validation and interpretation of FE results.

Chapter 4 adequately reviews essential and relevant theories of the mechanics of materials. Uniaxial stress–strain curve covering elastic-plastic response of solder materials is described. Effects of temperature and strain rate on mechanical properties and hardening behavior of materials are discussed. Cyclic behavior and creep response of the material are discussed in terms of relevant numerical models. Of particular interest is the unified inelastic strain theory (Anand model) for lead-free solder alloys. Relevant properties and behavior of tin-based solder alloys are presented.

Chapters 5 and 6 illustrate the FE simulation methodology described above through case studies in (Case I) reflow cooling of a typical microelectronic assembly with BGA solder joints and (Case II) deformation of the BGA assembly under reliability temperature and mechanical load cycles. The latter includes case studies on BGA assembly subjected to cyclic flexural load and cyclic torsional load. The predicted evolution of internal states of displacements, strains, and stresses in critical solder joints are presented and discussed with respect to the solder joint reliability.

Chapter 7 dwells on the concept of materials damage and damage mechanics approach in predicting deformation and failure processes of solder joints in the assembly. The concept of cohesive behavior of solder/IMC interface is introduced to simulate fracture process of the interface. Continuum damage model is introduced for predicting the progressive failure process of the bulk solder joint. Extension of existing cohesive zone model formulation to accommodate cyclic loading is discussed.

Predictive capability of these damage-based models for simulation of fracture processes of solder joints is illustrated in case studies of different loading scenarios, as elaborated in Chaps. 8 and 9. Case III describes the dynamic fracture process of solder/IMC interface during board-level drop test with a BGA assembly. Case IV illustrates the characteristic incremental damage evolution and fracture process in a solder joint under fatigue loading. Competition between solder/IMC interface damage and continuum damage in the bulk solder joint is examined through a case study on an assembly with a single solder joint subjected to cyclic shear-dominated loading. Simulation of failure of solder joints in BGA assembly subjected to flexural loading is also illustrated.

A short closure on the topic is covered in the last chapter.

Acknowledgments

The authors acknowledge the continuous financial support from Intel Technology (Malaysia) since year 2003 through *Intel Research Grants* for a series of research projects under UTM-Intel Contract Research program. Financial support from the Ministry of Science, Technology and Innovation (MOSTI) Malaysia through *ScienceFund Research Grant (2007–2009)* is acknowledged. These projects cover the challenging scope on the development of a validated methodology for reliability prediction of solder joints in BGA packages and assemblies. Dr. Loh Wei Keat of Intel Technology (Malaysia) served as the industry mentor for the program.

Many thanks are extended to research students involved in these projects, who have generated new knowledge and numerous refinements in simulation of solder materials and reliability of BGA packages. The outcomes of your countless hours of research are proudly included in this writing:
Koh Yee Kan
Liew Yek Ban
Ng Chee Weng
Lai Zheng Bo
Fethma M. Nor
Alif Farhan Mohd Yamin

A note of thanks is also given to Assoc. Prof. Dr. Nazri Kamsah and Prof. Dr. Ali Ourdjini, who have contributed toward successful and timely completion of our research projects and supervision of research students at the Computational Solid Mechanics Laboratory (CSMLab), UTM.

Also, thanks to Ms. Marhaini Mohamad Ibrahim for helping us with the preparation of the manuscript and to Mr. Mohd Hasri Mohd Harizan (Intel Technology) for effective coordination of the university–industry research program.

Contents

1 Introduction 1
1.1 Trends in Electronics' Packaging 1
1.2 Electronic Assembly with Flip Chip Package 2
1.3 Solder Joint Reliability Assessment 3
1.4 Solder Joint Reliability Simulation 5
References 6

2 Overview of the Simulation Methodology 7
2.1 Introduction 7
2.2 Simulation of Physical Phenomena 7
2.3 Numerical Experiment Framework 9
2.4 Deriving Finite Element Equations 12
2.4.1 Variation of Field Variables 13
2.4.2 Constitutive Equations 16
2.4.3 Total Potential of the System: The Functional 16
2.5 Procedures for Finite Element Simulation 17
References 21

3 Essentials for Finite Element Simulation 23
3.1 Introduction 23
3.2 Model Geometry 24
3.3 Finite Element Mesh and Mesh Study 26
3.4 Boundary Conditions 28
3.5 Load and Load Cycles 29
3.6 Material Models 33
3.7 Aspects of Model Validation 37
3.8 Interpretation of Results 39
References 43

4 Mechanics of Solder Materials 45
4.1 Introduction 45
4.1.1 Metallurgical Aspects of Solder Alloys 46

4.2 Elastic Behavior and Yielding ... 49
4.2.1 Generalized Hooke's Law ... 50
4.2.2 Thermal Strains ... 52
4.2.3 Yield Criteria ... 54
4.3 Plasticity and Hardening ... 56
4.4 Creep Response ... 60
4.5 Unified Inelastic Strain Theory ... 63
4.6 Cyclic Behavior ... 68
References ... 70

5 Application I: Solder Joint Reflow Process ... 75
5.1 Introduction ... 75
5.2 Finite Element Modeling ... 76
5.2.1 Constitutive Model for SAC405 Solder ... 78
5.3 Residual Internal States in the Critical Solder Joint ... 80
5.3.1 Reflow Process-induced Warpage of the Assembly ... 80
5.3.2 Evolution of Residual Stresses and Strains ... 83
5.3.3 Residual Stress and Strain Fields ... 84
References ... 85

6 Application II: Solder Joints Under Temperature and Mechanical Load Cycles ... 87
6.1 Introduction ... 87
6.2 BGA Assembly Under Temperature Cycles ... 88
6.2.1 Finite Element Modeling ... 88
6.2.2 Evolution of Internal States in the Solder Joint ... 89
6.2.3 Stress-Strain Hysteresis ... 92
6.2.4 Fatigue Life Prediction Models ... 94
6.3 BGA Assembly Under Cyclic Flexural Load ... 97
6.3.1 Finite Element Modeling ... 97
6.3.2 Deformation of the BGA Test Assembly ... 101
6.3.3 Evolution of Internal States in the Solder Joint ... 102
6.3.4 Solder Joint Fatigue Life Models ... 105
6.4 BGA Assembly Under Cyclic Torsional Load ... 106
6.4.1 Finite Element Modeling ... 106
6.4.2 Evolution of Stress and Inelastic Strain in the Critical Solder Joint ... 109
References ... 115

7 Damage Mechanics-Based Models ... 117
7.1 Introduction ... 117
7.2 Material Damage Concept ... 119
7.2.1 Mechanical Representation of Damage ... 120

7.3 Cohesive Zone Model. 122
7.3.1 Cohesive Behavior 123
7.3.2 Key Equations for Cohesive Zone Model 124
7.3.3 Determination of Model Parameter Values 127
7.3.4 Extension of Cohesive Model for Cyclic Loading. 129
7.4 Continuum Damage Model 133
7.4.1 Cyclic Continuum Damage Model. 133
References 134

8 Application III: Board-Level Drop Test 137
8.1 Introduction 137
8.2 Simulation of Board-Level Drop Test 138
8.2.1 Mathematical Model. 139
8.3 Finite Element Modeling 142
8.4 Interface Damage Initiation and Dynamic Propagation. 144
8.4.1 Modal Analysis of the PCB Assembly 144
8.4.2 Deflection of the Test Board. 145
8.4.3 Stresses and Inelastic Strains in Critical Solder Joints. 146
8.4.4 Dynamic Damage Process of Solder/IMC Interface 147
8.4.5 Aspects of FE Model Validation. 150
References 151

9 Application IV: Fatigue Fracture Process of Solder Joints 153
9.1 Introduction 153
9.2 Cyclic Shear Loading of a Solder Joint 154
9.2.1 Mathematical Model. 154
9.3 Finite Element Modeling 157
9.4 Stress and Inelastic Strain Distribution in the Solder Joint 158
9.4.1 Stress Distribution in the Solder Joint 158
9.4.2 Accumulated Inelastic Strain in the Solder Joint 161
9.5 Characteristic Damage Evolution. 161
9.6 Cyclic Flexural Loading of BGA Assembly with Damaging Solder Joints 165
9.6.1 Finite Element Modeling 166
9.6.2 Distribution of Internal States in the Critical Solder Joint . . . 167
9.6.3 Evolution of Inelastic Strain and Propagation of Damage in Solder Joints 168
References 171

Closure 173

Chapter 1
Introduction

1.1 Trends in Electronics' Packaging

The rapid evolution of electronic products into smaller forms, with greater functionality and assured safety and reliability, has led to an accelerated development in electronic packaging. Electronic packaging refers to technologies associated with producing an enclosure for an interconnection of integrated circuits to form electronic systems. Simultaneous development of Surface Mount Technology (SMT) has facilitated the assembly process at board level for these packages. In SMT, the electronic package is mounted directly on printed wiring or the Printed Circuit Board (PCB). However, it became a bottleneck to high speed computing by the year 2000 when the advancement of packaging technology could not keep pace with semiconductor IC technology [4]. Over the years of computing hardware development, an exponential growth in the number of transistors per square inch on an integrated circuit has been observed, as explained by Moore's law [8]. This trend shows data density doubling approximately every 18 months, as depicted in Fig. 1.1, and is expected to hold for another two decades [7]. Higher transistor integration density is a possibility with 3D interconnect technology, which allows for vertical stacking of the layers of basic electronic components using through-silicon via (TSV) interconnection, as envisioned in the International Technology Roadmap for Semiconductors [1]. The increase in complexity of newly designed packages is compounded by the use of different types of new materials, higher induced stress levels, aggressive operating environments and stringent reliability requirements. In this regards, European legislations on Waste Electrical and Electronic Equipment (WEEE) and Restriction of the use of Hazardous Substances (RoHS) have seen electronics manufacturing industries shifting from leaded to lead-free solder alloys since 2006 [12, 17]. Although the National Electronics Manufacturing Initiative (NEMI) organization recommends the use of Sn–3.9Ag–0.6Cu alloy for the solder reflow process and Sn–0.7Cu alloy for wave soldering, different variations of tin–silver–copper (SAC) alloys, including SAC305, SAC387 and SAC405, have been employed [9]. The different

M. N. Tamin and N. M. Shaffiar, *Solder Joint Reliability Assessment*,
Advanced Structured Materials 37, DOI: 10.1007/978-3-319-00092-3_1,

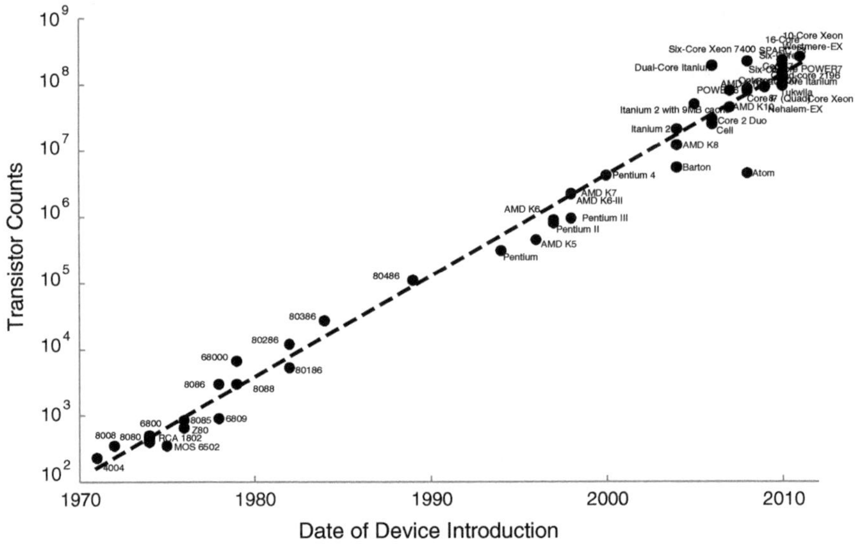

Fig. 1.1 Exponential growth of the number of transistors per square inch on an integrated circuit, as observed by Moore [8]

compositions and microstructures of these alloys necessitate costly and time-consuming experimental programs to determine properties and behaviors as part of the requirements in assessing and establishing solder joint reliability.

1.2 Electronic Assembly with Flip Chip Package

An electronic package consists of semiconductor devices, such as IC chips, connected to external circuitry using pre-deposited solder bumps on chip pads. In a flip chip package the solder bumps on the chip pads are deposited on the topside of the Si wafer during the final wafer processing step. The chip is flipped over and the solder bumps are aligned with the matching pads on the external circuitry during fabrication. The package is then reflowed to complete the interconnection. This flip chip method is also known as *Controlled Collapsed Chip Connection (C4)*. Flip chip technology has been applied in interconnecting a large number of I/Os and area array solder bumps on silicon chips with substrate because of its high interconnect speeds, high density and low profiles [5]. A typical cross section of a Flip Chip Ball Grid Array (FCBGA) package is schematically illustrated in Fig. 1.2a. The package consists of silicon die; a semiconductor material onto which integrated circuit is etched, substrate; a base plate onto which other materials are deposited, solder bumps providing interconnections between the silicon die and substrate, and underfill; an epoxy-based filler material providing strength and protection against moisture. A polymer-based over-mold is often applied to further

Fig. 1.2 **a** Schematic cross section of an electronic assembly with a flip chip package and **b** electronic packages on a substrate with BGA solder joints

protect the package from a harsh operating environment, as shown in Fig. 1.2b. The FCBGA package features a high pin count, high speed and good thermal dissipation when compared to other types of electronic packages [13].

The flip chip package is then connected to the printed wiring or circuit board (PCB) in the second level packaging to form an electronic assembly. Such interconnections can be designed as pin-and-socket joints for ease of removing the package, or as permanent mechanical joints using BGA solders, as illustrated in Fig. 1.2a. In the latter, the solder joints are formed following the solder reflow cooling process. The process exposes the assembly to temperature history with the peak temperature level corresponding to the melting point of the solder material. The large temperature difference during cooling from reflow to ambient temperature induces residual strains and stresses in various components of the assembly due to mismatches in the coefficient of thermal expansion of the different materials that make up the assembly. Temperature excursions during service further aggravate material degradation through the combined creep-fatigue-environmental effects and can cause premature failure of the assembly. Consequently, the reliability of the package and assembly during service conditions need to be quantified.

1.3 Solder Joint Reliability Assessment

Solder bumps or joints provide both electrical connections between the Si chip and the external circuitry on the substrate and PCB. The joints also ensure the mechanical strength and the stiffness of the package. Yet, the solder joint is often

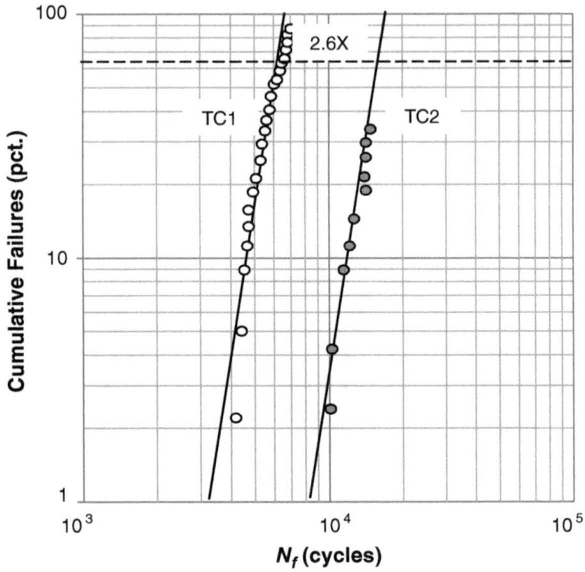

Fig. 1.3 Statistical distribution of fatigue lives for electronic packages at different accelerated reliability temperature cycles, illustrating the acceleration factor

the weakest element in a typical BGA package and assembly. Thus, the reliability of the device depends on the reliability of the solder joint. Reliability of the solder joint refers to the ability of the solder interconnect to retain functionality throughout the design life of the package or assembly. Premature failure of the solder joint is caused primarily by induced stresses in the joint due to mismatches in the coefficient of thermal expansions of the various package materials during the solder reflow cooling process and temperature excursions during operation of the device. Consequently, assessment of the solder joint failure process is paramount in ensuring the reliability of a device. Reliability of newly designed packages and assemblies is assessed by subjecting the components to temperature and/or mechanical stressing and establishing their lives. The reliability test programs often employed are accelerated stress testing by temperature cycles, mechanical load cycles, drop impact load, and combined temperature and humidity conditions. These reliability test procedures are described in testing standard documents (e.g. [2, 3]). Such reliability tests are targeted to produce identical failure mechanisms to those found during service conditions for the specific package design and generate relevant information for understanding the physics of package failures [16]. The failure process of solder joints is governed by several phenomena, including microstructure evolution, creep deformation, thermo- and electro-migration, fatigue damage accumulation and dynamic fracture. In addition, accelerated test data are collected to represent the statistics of failures [15]. The measured life data from these tests determines the mean time of failure for the package when subjected to a specified accelerated test condition. The correlation of these statistical data sets for response of packages to different loading conditions is expressed in terms of an acceleration factor, as illustrated in Fig. 1.3. Such acceleration factors are extrapolated for the

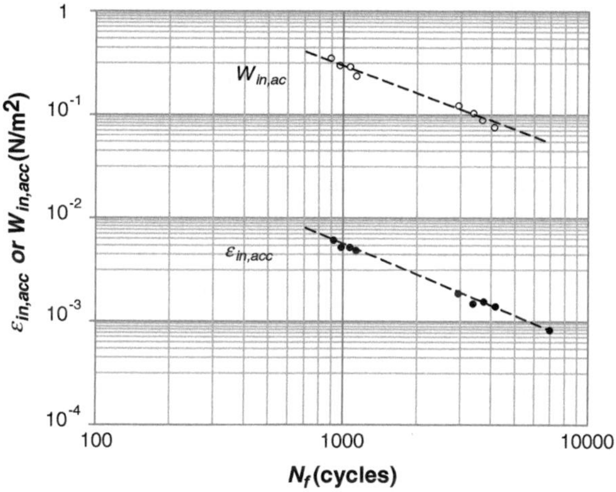

Fig. 1.4 Reliability model of electronic packages based on the measured fatigue life of solder joints and predicted characteristic fatigue variable of solder materials

operating conditions of similar packages. An extensive review of various empirical models available is presented elsewhere (e.g. [6]). A drawback of the phenomenological-based empirical models in a life prediction scheme is a lack of the underlying physical mechanisms of failure in solder joints. This limits the applicability of the model to identical packages and assemblies. While reliability test data provides a statistical life for packages, a failure analysis of these packages identifies dominant mechanisms of failure.

1.4 Solder Joint Reliability Simulation

A numerical simulation of solder joint reliability in the finite element method (FEM) complements the accelerated stress testing of microelectronic packages in the faster generation of reliability test data. The finite element (FE) simulation presents a systematic approach in predicting the useful life of the package by examining internal states of displacements, inelastic strains and stresses in the critical solder joint for different reliability loading of the package. Selected parameters, such as accumulated inelastic strain or plastic work density per load cycle of the solder joint, are commonly used to measure fatigue life of similar packages to establish the fatigue life model of the solder joints. A typical outcome of this experimental-computational approach is illustrated in Fig. 1.4. In order to accurately capture the mechanics and mechanisms of the failure process, the damage mechanics-based model has been examined. In this approach, the damage initiation event and subsequent evolution of damage representing material deterioration and crack propagation in the critical solder joint are tracked up to the final separation of the material point. It is worth noting that different package designs, distinguished by different geometry and the array of solder joints lead to different stress fields and concentrations, and possibly different fatigue lives, thus the reliability of the packages. In addition, new materials call for a new set of properties and constitutive behavior to be established through the expensive testing program.

FE simulation is commonly employed in addressing issues of solder joint reliability as an integral part of the Design-for-Reliability approach [10, 11, 14]. The challenge is to generate solutions of known accuracy and validity within pre-determined application limits. The simulation could result in shortening product development time, increasing product reliability and providing a better understanding of product design limits.

References

1. International Technology Roadmap for Semiconductors (ITRS) (2012). www.itrs.net
2. JESD22–B111.: Board Level Drop Test Method of Components for Handheld Electronic Products. JEDEC Solid State Technology Association, VA, USA (2003)
3. JESD22–B104C: Mechanical shock. JEDEC Solid State Technology Association, VA, USA (2004)
4. Lau, J.H. (ed.): Low Cost Flip Chip Technologies: For DCA, WLCSP, and PBGA Assemblies. McGraw-Hill, New York (2000)
5. Lau, J.H.: Flip Chip Technologies. International Technology Roadmap for Semiconductors (ITRS). McGraw-Hill, New York (1996)
6. Lee, I., Lee, W.W., Nguyen, L.T., Selvaduray, G.S.: Solder joint fatigue models: review and applicability to chip scale packages. Microelectron. Reliab. **40**, 231–244 (2000)
7. Mahajan, R., Nair, R., Wakharkar, V., Swan, J., Tang, J., Vandentop, G.: Emerging directions for packaging technologies. Intel Technol. J. **6**(2), 62–75 (2002)
8. Moore, G.E.: Cramming more components onto integrated circuits. Electron Mag. 38(8), pp. 114 (1965)
9. National Electronics Manufacturing Initiative: Roadmap of Lead-free Assembly in North America (NEMI) (2002). www.inemi.org
10. Pang, J.H.L., Low, T.H., Xiong, B.S., Che, F.X.: Design for reliability (DFR) methodology for electronic packaging assemblies. In: Proceedings of 5th Electronics Packaging Technology Conference, pp. 470–478 (2003)
11. Pang, J.H.L.: Lead-free solder materials: design for reliability. In: Micro- and Opto-Electronic Materials and Structures: Physics, Mechanics, Design, Reliability, Packaging. Springer, New York. vol. 1, pp. 429–456 (2007)
12. RoHS, European Parliament: Directive 2002/95/EC on the Restriction of the Use of Certain Hazardous Substances in Electrical and Electronic Equipment. Official J Eur. Union. L37/19–L37/23 (2003)
13. Sawada, Y., Harada, K., Fujioka, H.: Study of package warp behavior for high-performance flip-chip BGA. Microelectron. Reliab. **43**(3), 465–471 (2003)
14. Sitaraman, S.K., Pang, J.H.L.: Fundamentals of design for reliability. In: Tumala, R.R. (ed.) Fundamentals of Microsystems Packaging. McGraw-Hill, New York (2001)
15. Suhir, E.: Accelerated life testing (ALT) in microelectronics and photonics: its role, attributes, challenges, pitfalls, and interaction with qualification tests1. J. Electron. Packag. **124**, 281–291 (2002)
16. Tummala, R.R.: Fundamentals of Microsystems Packaging. McGraw Hill, New York (2001)
17. WEEE, European Parliament: Directive 2002/96/EC on Waste of Electrical and Electronic Equipment. Official J Eur. Union. L37/24–L37/38 (2003)

Chapter 2
Overview of the Simulation Methodology

2.1 Introduction

This chapter presents an overview of the simulation methodology that comprises both the art and science involved in simulating physical phenomena. It adequately summarizes the various aspects of simulation, including identification of the physical problem of interest, determination of material properties and behavior through mechanical testing, and formulations of governing equations for the finite element (FE) method. The chapter also introduces a numerical experiment framework along with a description of a typical problem solving process using FE simulation. In addition, it elaborates generic steps in performing FE simulation of structural engineering problems. Examples related to assessment of solder joint reliability are illustrated.

2.2 Simulation of Physical Phenomena

The central problem being addressed throughout this book relates to the assessment of solder joint reliability in microelectronic packages and assemblies. A quantitative assessment of reliability covers the establishment of the mechanics of solder joints under temperature and mechanical cycles, a description of the fracture process in critical solder joints and a determination of the fatigue lives of the solder joints. The small physical dimensions of solder joints found in a typical BGA package render direct measurements of parameters and property values of the solder joint difficult if not impossible. In this respect, the hybrid experimental-computational approach offers an indirect assessment of package reliability. Such methodology calls for the simulation of various phenomena exhibited by solder joints in the package and assembly during reliability testing and field operation.

Simulation is a process of recreating physically occurring phenomena of interest for a better understanding of the behavior of the system and materials, or for

M. N. Tamin and N. M. Shaffiar, *Solder Joint Reliability Assessment*,
Advanced Structured Materials 37, DOI: 10.1007/978-3-319-00092-3_2,

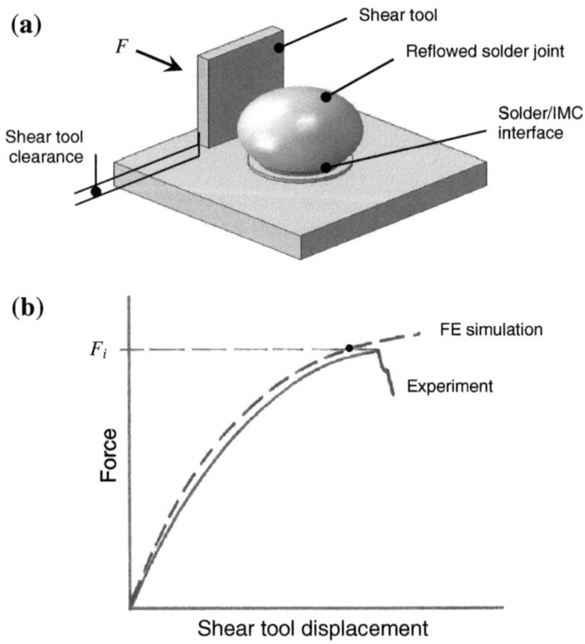

Fig. 2.1 Experimental-computational concept for the determination of solder/IMC interface shear strength. **a** Solder ball shear push test set up. **b** Comparison of FE-predicted and measured force-displacement curves

establishing their characteristics. Examples of these phenomena are viscoplastic deformation of metals, creep-fatigue interaction, brittle fracture of the bi-material interface, convective heat transfer and fluid-structure interaction. Simulation of any of these phenomena for a specific problem, such as fatigue failure of solder joints in a BGA package, is often accomplished using combinations of experimental characterization, mathematical equations and/or computational methods. An example of the experimental-computational concept involving a simulation of a solder ball shear push test of a reflowed solder joint specimen is illustrated in Fig. 2.1 and described in the following paragraphs.

The solder ball shear push test is a common test performed on reflowed solder specimens to partly quantify intrinsic properties of solder/intermetallic (IMC) interfaces. It is challenging to fabricate a standard shear test specimen with similar solder/IMC interface in the gage section as one found in reflowed solder joints. Such solder/IMC interface is a result of a chemical reaction between the solder and *Ball Limiting Metallurgy* (*BLM*) layers of the substrate during the solder reflow process, thus forming a thin intermetallic layer. Since the type of the intermetallic phase and grain structure of the adjacent reflowed solder dictate the resulting strength properties of the interface, the solder specimen is prepared using a similar reflow process to that of the package with BGA solder joints.

The experimental part of the hybrid experimental-computational approach consists of the solder ball shear push test. The test is performed on a reflowed solder joint specimen under displacement-controlled conditions. The test setup is illustrated in Fig. 2.1a. The resisting force and shear tool displacement data pairs are recorded throughout the test. A schematic of a measured force-displacement curve for a solder joint with a relatively brittle solder/IMC interface is represented in Fig. 2.1b by the solid line. A sudden drop in the force value at the peak of the curve indicates the onset of solder/IMC interface fracture. The corresponding stress in a critical point on the solder/IMC interface should have reached the shear strength of the interface. The nominal shear strength of the interface can be defined as the recorded peak force, F_i divided by the measured sheared area of the solder joint. However, it is argued that the inherent shear tool clearance in the test setup induces significant bending stress on the interface, thus the measured force constitutes the combined effects of shear and bending stresses on the interface [8].

An FE simulation of the shear push test is then employed to gain insight into the stress states at the critical point on the interface corresponding to the measured peak force. An accurate FE model of the shear push test setup is designed, as explained in Chap. 3. Comparable force-displacement curves, as predicted by the FE model (illustrated by the dashed line in Fig. 2.1b) and that measured experimentally should be ensured. Equivalent shear stress at the interface is represented, in this example, by the absolute maximum shear stress quantity. The typical distribution of the absolute maximum shear stress on Sn-40Pb solder/Ni_3Sn_4 interface corresponding to the measured peak force is illustrated in Fig. 2.2. The stress field exhibits a gradient of the absolute maximum shear stress values at the leading edge of the interface. The highest magnitude of the absolute maximum shear stress at the edge of the solder/IMC interface is taken as the shear strength of Sn-40Pb solder/Ni_3Sn_4 bi-material interface at the test temperature. It is noted that the nominal shear strength calculated based on the observed peak force over the sheared area corresponds to the predicted minimum magnitude of the absolute maximum shear stress in the central portion of the interface plane [8]. Fractographic analysis of the fractured interface should reveal that shear failure initiated at the solder/IMC interface. Such observation provides validation of the interface shear strength determined using the experimental-computational approach.

A similar concept of combining experimental methods with a computational approach by FE simulation is employed in various studies in conjunction with the assessment of solder joint reliability. This hybrid experimental-computational approach is incorporated within a numerical experiment framework.

2.3 Numerical Experiment Framework

The numerical experiment framework encompasses several requirements in addressing an engineering problem. The framework is schematically illustrated in Fig. 2.3. Emphasis is placed on the complete identification and synthesis of the real-world problem of interest. Utilization of computational tools such as FE analysis software and spreadsheets are central to the problem solving process. Oftentimes, controlled experiments are required both in establishing parameter values of the material model and validating predicted results.

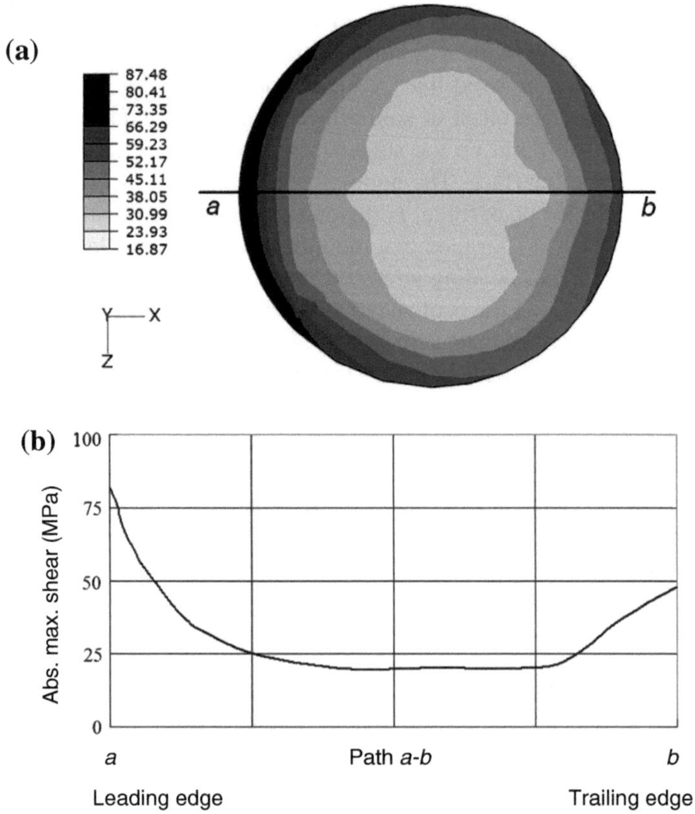

Fig. 2.2 **a** Distribution of the absolute maximum shear stress and **b** Variation of the absolute maximum shear stress along *path a–b* on the solder/IMC interface plane

The solder joint reliability assessment is used as an example of the real-world problem in describing the numerical experiment framework, shown in Fig. 2.3. Synthesis of the problem calls for a thorough understanding of the failure process of solder joints in a BGA package during reliability temperature cycles. Reliability temperature cycles range typically from −40 to 125 °C along with relatively high heating and cooling ramps. The high homologous test temperature activates creep deformation of the solder, especially when dwell time period at a peak temperature level is incorporated into the cycle. The large temperature range induces plasticity through mismatches in the coefficients of thermal expansion among the various types of materials making up the assembly. The fast temperature ramp rates induce viscoplastic deformation while the temperature cycles promote fatigue and creep-fatigue interaction of the solder joints. These mechanical responses should be represented by appropriate constitutive models for the stress–strain behavior of a material point under loading. Damage-based models may be required for fracture prediction of the solder joints.

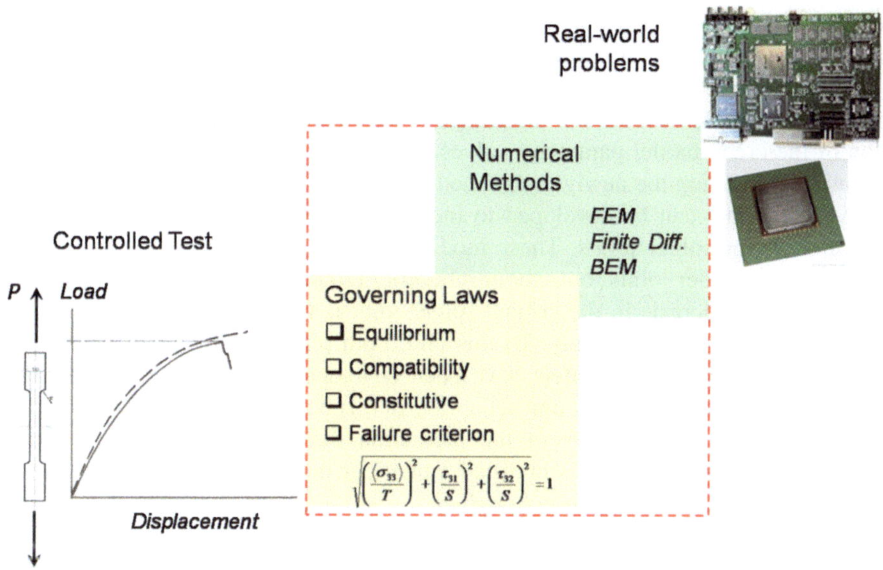

Fig. 2.3 A framework for numerical experiment employing the experimental-computational approach

The solution approach should integrate various physical laws including equilibrium of forces, strain-displacement compatibility conditions, minimum potential energy requirements and plastic flow rule in the solution process. These governing laws have been formulated into numerical procedures such as the finite element method (FEM), the finite difference method and the boundary element method. While details of these methods are beyond the scope of this book, essential features of the FEM will be discussed in relation to the accuracy of computation, validity of prediction and limitation of the solution.

The choice of suitable and valid constitutive models is central in ensuring accurate prediction of the stress–strain response of the solder joint material to load. In view of the anticipated temperature- and strain rate-dependent response of the solder alloy, the viscoplastic constitutive model, the creep and the fatigue model of the material can be employed in the simulation. Several constitutive models such as the Anand [2] and the Johnson-Cook [6], and the creep models, including hyperbolic sine and Arrhenius equations, have been coded in commercial finite element analysis (FEA) software. In addition, continuum damage-based material models are useful in simulating the fatigue failure process in bulk solder joints. The challenge in employing any of these material models lies in determining the values of the model parameters over the temperature and strain range of interest.

Values for the material model parameters need to be extracted from test data for the material under controlled experimental conditions. A comprehensive experimental program is required to establish uniaxial stress–strain diagrams for the solder alloy at different test temperature levels and applied strain rates. These curves are then used to establish values of the set of constitutive model parameters for the solder alloy. An additional discussion on extracting model parameter values from test data is included in Chap. 4. Similarly, a series of creep-rupture curves at combinations of applied stress and test temperature levels are required for determination of the creep model parameter values. An additional set of test data is needed for use in validating the newly established model.

An FE model can be developed to incorporate relevant constitutive and damage models for solder joints. These models predict the evolution of stresses and strains in the solder joints throughout reliability temperature cycles. Characteristic stress–strain hysteresis in the critical solder joint is then employed in the fatigue life prediction of the assembly. To this end, both phenomenological- and mechanism-based fatigue life models have been developed. Selected models are discussed with respect to solder joint reliability in this book.

The accuracy and validity of the prediction through numerical experiments rely on the capability of the simplified numerical model employed, the accuracy of the input parameters and good engineering judgment on the calculated results. Challenges in employing numerical experiments in the evolution of technology are derived from the requirements to realistically mimic physical phenomena. These include modeling the compounding effect and the various effects of microstructure features, the long-term response and degradation of materials, and the processing-structure-property relationship. The success of numerical experiments strongly depends on the solid foundation of physics, analytical capability, and the added advantage of parallel computing capabilities on multi-core, multi-processor computers.

2.4 Deriving Finite Element Equations

The finite element method (FEM) is a numerical analysis technique for obtaining approximate solutions to a wide variety of engineering problems. The basic premise of the method is that a solution region can be analytically modeled or approximated by replacing it with an assemblage of discrete elements, as illustrated in Fig. 2.4. Within each element, the field or dependent variables of interest such as displacement, temperature and damage are governed by relevant differential equations, while specified boundary conditions are satisfied at the boundary of the solution region. Extensive treatment of FEM is beyond the scope of this book and covered in numerous publications (e.g. [3–5, 7]). This section provides a basic understanding of FEM by examining how finite element equations are derived from problems in solid mechanics.

The formulation of the displacement-based finite element (FE) method can be based on several approaches, including the variation method, the use of the principal of virtual displacement, the Galerkin method and the Ritz method. The following derivation of FE equations is illustrated using the variational approach in establishing the governing equilibrium equations of a system. The requirements

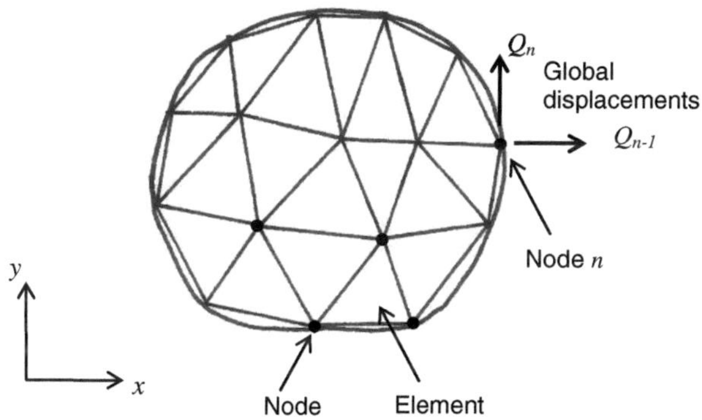

Fig. 2.4 Discretization of a solution domain into assemblage of finite elements

and assumptions for deriving the element equations are discussed in the following. The basic 3-node triangular element along with the linear elastic material response is used in this illustration, which aims at providing a quick overview of the steps in deriving FE equations. The element equations are dedicated for plane stress and plane strain problems in solid mechanics.

FE equations for this element are expressed in the matrix form as:

$$[k]\{q\} = \{f\} \tag{2.1}$$

where $[k]$ is the stiffness matrix, vector $\{q\}$ consists of nodal displacement values and vector $\{f\}$ is the corresponding element nodal forces. Derivation of the FE equations begins with the assumed displacement field within an element. One such element is shown in Fig. 2.5 along with nodal coordinates and displacement variables. The displacement component in the x- and y-axis is denoted as u and v respectively.

2.4.1 Variation of Field Variables

Displacements within the element are assumed to vary according to the prescribed interpolation or shape functions, N_i ($i = 1, 2, 3$) such that:

$$\{u\} = \begin{bmatrix} N_1 & 0 & N_2 & 0 & N_3 & 0 \\ 0 & N_1 & 0 & N_2 & 0 & N_3 \end{bmatrix} \begin{Bmatrix} q_1 \\ q_2 \\ q_3 \\ q_4 \\ q_5 \\ q_6 \end{Bmatrix} \tag{2.2}$$

The shape functions for the 3-node triangular element can be expressed in terms of Cartesian coordinates of the element nodes:

$$N_i = \frac{1}{2\Delta}(a_i + b_i x + c_i y), \quad i = 1, 2, 3 \tag{2.3}$$

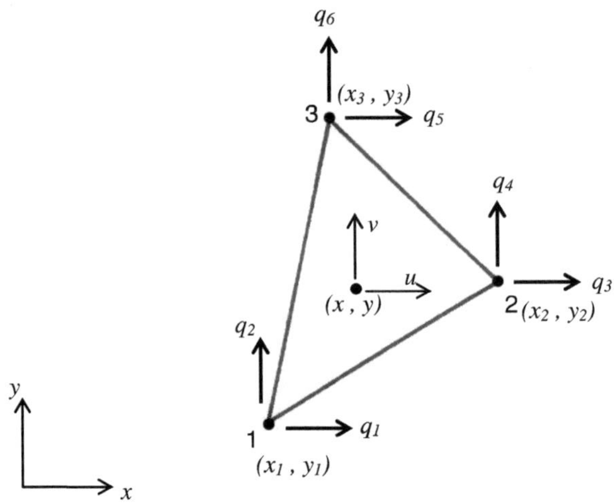

$$a_1 = x_2y_3 - x_3y_2\,,\quad a_2 = x_3y_1 - x_1y_3\,,\quad a_3 = x_1y_2 - x_2y_1$$
$$b_1 = y_2 - y_3\,,\quad b_2 = y_3 - y_1\,,\quad b_3 = y_1 - y_2$$
$$c_1 = x_3 - x_2\,,\quad c_2 = x_1 - x_3\,,\quad c_3 = x_2 - x_1$$

$$A = \frac{1}{2}\det\left|[J]\right| \quad ; \quad [J] = \begin{bmatrix} (x_1 - x_3) & (y_1 - y_3) \\ (x_2 - x_3) & (y_2 - y_3) \end{bmatrix}$$

$$\{q\} = \begin{Bmatrix} q_1 \\ q_2 \\ q_3 \\ q_4 \\ q_5 \\ q_6 \end{Bmatrix} = \begin{Bmatrix} u_1 \\ v_1 \\ u_2 \\ v_2 \\ u_3 \\ v_3 \end{Bmatrix}$$

Fig. 2.5 A 3-node triangular element shown with element nodal displacements. An element nodal connectivity is defined to provide the connection between local and global node numbers for each element

Indices i in Eq. (2.3) refer to the assigned node number for the element while Δ represents the area of the element that could also be computed using nodal coordinates of the element (refer to Fig. 2.5). In addition, the coefficients, a_i, b_i and c_i are also functions of nodal coordinates of the element, as included in Fig. 2.5. Requirements for the choice of the shape functions including continuity and compatibility conditions are deliberated elsewhere (e.g. [3, 4]).

Alternate forms of the shape functions can also be prescribed using generalized or local coordinates (ξ, η) as:

$$\begin{aligned} N_1 &= \xi \\ N_2 &= \eta \\ N_3 &= 1 - \xi - \eta \end{aligned} \tag{2.4}$$

Both coordinates values vary, such that $0 \le (\xi, \eta) \le 1$. The shape functions in Eq. (2.4) should also satisfy the condition $N_1 + N_2 + N_3 = 1$. Each shape function takes a value of unity at the respective node and varies linearly to zero at the other two nodes. Schematic representations of the shape functions are shown in Fig. 2.6.

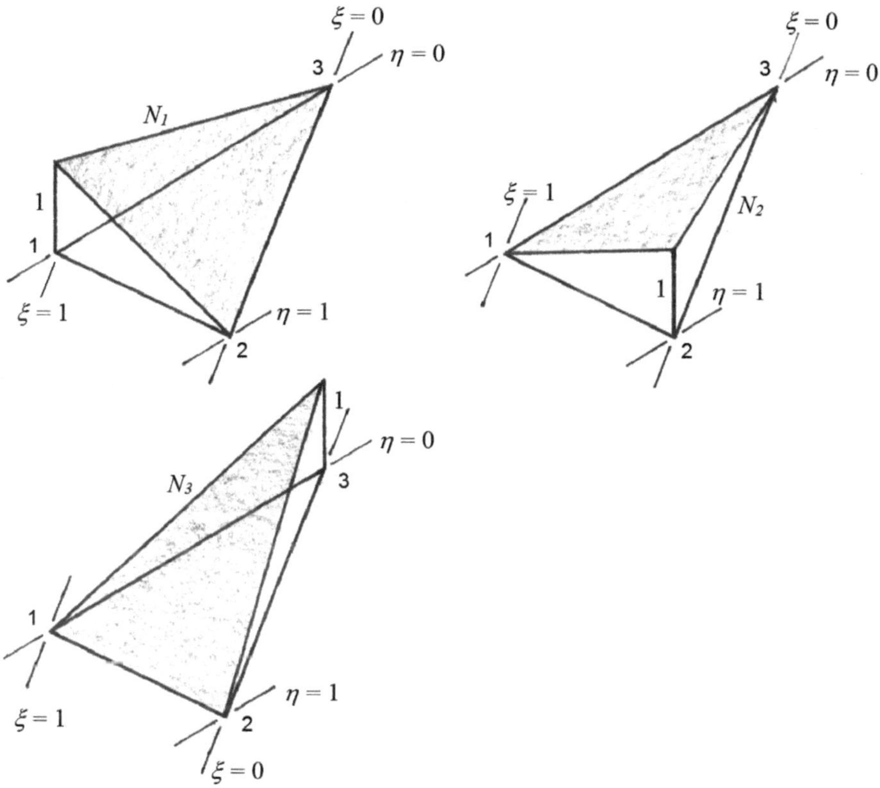

Fig. 2.6 The representation of shape functions in the generalized (local) coordinate system for a 3-node triangular element

The strain-displacement relationship or compatibility conditions, based on small strain theory, can be expressed as:

$$\{\varepsilon\} = \begin{Bmatrix} \varepsilon_x \\ \varepsilon_y \\ \gamma_{xy} \end{Bmatrix} = \begin{Bmatrix} \frac{\partial u}{\partial x} \\ \frac{\partial v}{\partial y} \\ \frac{\partial v}{\partial x} + \frac{\partial u}{\partial y} \end{Bmatrix} = \begin{bmatrix} \frac{\partial N_1}{\partial x} & 0 & \frac{\partial N_2}{\partial x} & 0 & \frac{\partial N_3}{\partial x} & 0 \\ 0 & \frac{\partial N_1}{\partial x} & 0 & \frac{\partial N_2}{\partial x} & 0 & \frac{\partial N_3}{\partial x} \\ \frac{\partial N_1}{\partial x} & \frac{\partial N_1}{\partial x} & \frac{\partial N_2}{\partial x} & \frac{\partial N_2}{\partial x} & \frac{\partial N_3}{\partial x} & \frac{\partial N_3}{\partial x} \end{bmatrix} \begin{Bmatrix} u_1 \\ v_1 \\ u_2 \\ v_2 \\ u_3 \\ v_3 \end{Bmatrix} \tag{2.5}$$

It is noted that only three independent strain components are included in the strain vector, $\{\varepsilon\}$ and sufficient for the plane elasticity problem. The last equality follows from the form of the displacement field, as defined in Eq. (2.2). Partial differentiation of Eq. (2.3) with respect to the respective coordinate variable x and y results in the following:

$$\frac{\partial N_i}{\partial x} = \frac{b_i}{2\Delta}; \quad \frac{\partial N_i}{\partial y} = \frac{c_i}{2\Delta}; \quad i = 1,\ 2,\ 3 \tag{2.6}$$

Thus, the strain-displacement relationships can be expressed in terms of nodal coordinate as:

$$\begin{Bmatrix} \varepsilon_x \\ \varepsilon_y \\ \gamma_{xy} \end{Bmatrix} = \begin{bmatrix} (y_2 - y_3) & 0 & (y_3 - y_1) & 0 & (y_1 - y_2) & 0 \\ 0 & (x_3 - x_2) & 0 & (x_1 - x_3) & 0 & (x_2 - x_1) \\ (x_3 - x_2) & (y_2 - y_3) & (x_1 - x_3) & (y_3 - y_1) & (x_2 - x_1) & (y_1 - y_2) \end{bmatrix} \begin{Bmatrix} u_1 \\ v_1 \\ u_2 \\ v_2 \\ u_3 \\ v_3 \end{Bmatrix} \tag{2.7a}$$

or alternately:

$$\{\varepsilon\} = [B]\{q\} \tag{2.7b}$$

The resulting strain-displacement matrix, $[B]$ consists only of nodal coordinate values for the element, which leads to a constant strain field within the element. Thus the element is also known as a constant-strain-triangular (CST) element.

2.4.2 Constitutive Equations

The constitutive equations for linear elasticity of numerous metallic materials are given by the generalized Hooke's law:

$$\{\sigma\} = [C]\{\varepsilon\} \tag{2.8}$$

The Cartesian stress and strain components are as defined in Chap. 4 [Eqs. (4.1a), (4.1b) and (4.2)]. The general form of the elasticity matrix $[C]$ is presented in Eq. (4.1a), while the specific form for the idealized plane stress or plane strain condition is shown by Eqs. (4.4) and (4.6), respectively.

2.4.3 Total Potential of the System: The Functional

The variational approach is employed in this illustration to derive the governing equilibrium equations, i.e. the FE equations of the continuous system. In this approach, the total potential, Π of the system is calculated and the stationary of Π is invoked with respect to the state variables. The total potential or *functional* of the problem in solid mechanics is represented by the total potential energy of the solid:

$$\Pi(u, v) = U(u, v) - W(u, v) \tag{2.9}$$

where $U(u, v)$ is the strain energy of the system and $W(u, v)$ is the potential of the applied loads. Considering elastic deformation of the solid (with area, A and thickness, *t*) under the acting body force, $\{F_b\}$ and applied traction, $\{T\}$ along the edge length, S_1 Eq. (2.9) can be written as:

$$\Pi(u,v) = \frac{1}{2}\iint\limits_{A} \left[\lfloor q \rfloor [B]^T [C][B]\{q\}\right] tdA - \iint\limits_{A} \lfloor F_b \rfloor \{q\} tdA - \int\limits_{S_1} \lfloor T \rfloor \{q\} dS \tag{2.10}$$

In invoking the principle that the total potential of the system must be stationary i.e. the variation, $\delta\Pi = 0$ implies the total potential energy of the deforming solid must be at a minimum for equilibrium conditions. The total potential energy for a discretized elastic domain is the sum of energies from all elements in the domain. Thus the minimum potential energy theorem requires that the displacement field $q(u, v)$ that satisfies the equilibrium of forces and the conditions at the boundary surface also minimizes the total potential energy of the system. Such requirements can be written as:

$$\begin{aligned} \delta\Pi(u,v) &= \sum_{e=1}^{M} \delta\Pi^{(e)}(u,v) = 0 \\ \delta\Pi^{(e)}(u,v) &= \sum_{i-1}^{3} \frac{\partial\Pi^{(e)}}{\partial u_i}\delta u_i + \sum_{i=1}^{3} \frac{\partial\Pi^{(e)}}{\partial v_i}\delta v_i = 0 \end{aligned} \tag{2.11}$$

Since δu_i and δv_i are independent variations and not necessarily zero, thus:

$$\frac{\partial\Pi^{(e)}}{\partial\{q\}} = \begin{Bmatrix} \frac{\partial\Pi^{(e)}}{\partial u_i} \\ \frac{\partial\Pi^{(e)}}{\partial v_i} \end{Bmatrix} = 0; \quad i = 1,\ 2,\ 3 \tag{2.12}$$

Operating on Eq. (2.10) yields the FE equations for the element (the superscript $^{(e)}$ has been omitted for convenience):

$$\iint\limits_{A} [B]^T [C][B] tdA\{q\} = \iint\limits_{A} N_i\{F_b\} tdA + \int\limits_{S_1} N_i\{T\} dS \tag{2.13a}$$

which can be expressed in the familiar form shown in Eq. (2.1).

2.5 Procedures for Finite Element Simulation

Finite element (FE) simulation refers to a series of related problem solving activities; while employing FEM for solving the governing equations to arrive at the solution of the identified engineering problem. Figure 2.7 illustrates the process flow in the simulation of physical problems employing FEM as a computational tool. FE simulation consists of a mathematical modeling phase, an FE modeling and solution phase, and an analysis phase. As part of problem identification, mathematical modeling scopes the complex physical problem with adequate details

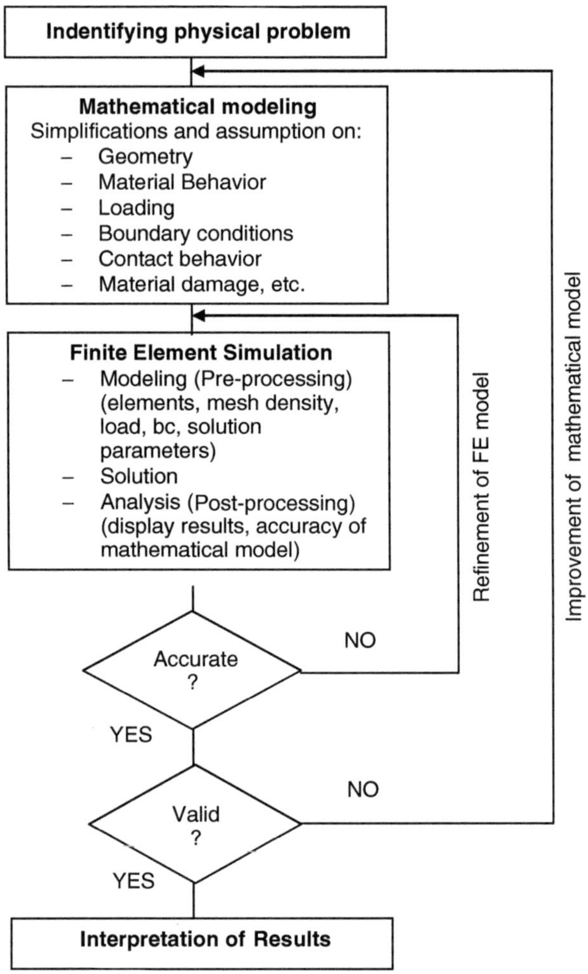

Fig. 2.7 Process flow of problem solving by finite element simulation

for FE simulation. These simplifications, while ensuring a correct and accurate predicted response of the model, limit the application of calculated results with respect to the various assumptions employed. In considering the relatively complex phenomenon of interacting surfaces, for example, a linear Coulomb's friction model is commonly employed. The assumption of an analytically *rigid body* is often used to model parts with high stiffness relative to the adjacent softer deformable part of interest. Such *rigid body* assumption reduces computational time and cost during the solution phase through reduction of active degrees of freedom. Other simplifications and assumptions on applied loading, material laws and boundary conditions are discussed with respect to specific examples in Chaps. 5, 6, 8 and 9.

Table 2.1 Steps in FE simulation (modeling, solution and post-processing phase) and the corresponding procedures performed by FEA software

FE procedures	FE software user steps
• Select the solution domain	• Draw the model geometry
• Discretize the continuum	• Mesh the model geometry
• Choose interpolation functions	• Select element type
• Derive element characteristics matrices and vectors	• (The FEA software was written to do this) Input material properties
• Assemble element characteristic matrices and vectors	• (The software will assemble them) Input specified load and boundary conditions
• Solve the system equations	• (The compiler will solve them) Request output
• Make additional computations, if desired	• Post-process the result files

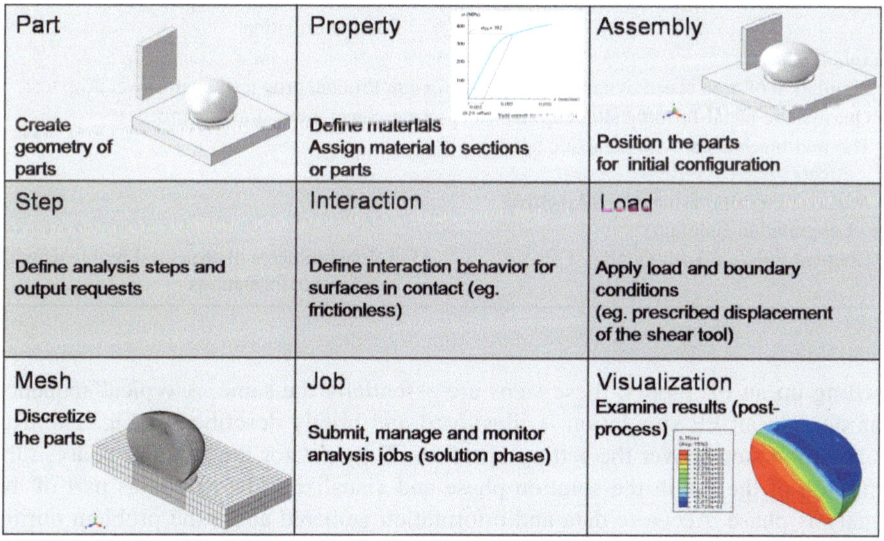

Fig. 2.8 The typical sequence of steps in FE simulation—setting-up an FE model, job submission for solution and visualization in the post-processing phase. Inset figures illustrate the simulation of reflowed solder ball shear push test. (Based on Abaqus FEA software (Abaqus [1])

The FE modeling phase is described on the assumption that one is using the commercially available finite element analysis (FEA) software. The FE modeling phase refers to the process of setting-up a geometrical model, specifying initial and boundary conditions, and prescribing load and load cases for the analysis, as required by the software. This phase is also called a pre-processing phase. The essential steps in FE modeling and the corresponding procedures performed by the pre-processor of FEA software are listed in Table 2.1. Although different commercial FEA software provides different *Graphical User Interfaces* (*GUI*) for

Table 2.2 The types of problems for implicit and explicit analysis in FE simulation

Implicit analysis	
• Static stress/displacement analysis	• Rate independent/dependent response
	• Eigenvalue buckling load prediction
• Linear dynamics	• Natural frequency extraction
	• Modal superposition
	• Response spectrum analysis
	• Random loading
• Linear/nonlinear dynamics	• Implicit/explicit transient dynamics
• Heat transfers/acoustics, mass diffusion, steady-state transport problems	
• Multiphysics analysis	• Thermal-mechanical analysis
	• Structural-acoustic
	• Fluid-structure interaction
	• Fully/partially saturated pore fluid flow-deformation
Explicit analysis	
• Simulation of high speed dynamic events	• e.g. Product drop test, board-level drop test
• Quasi-static metal-forming simulations	• e.g. Sheet metal drawing process
• Thermal-mechanical with adiabatic heating effects	
• Adaptive meshing using ALE (Adaptive Langrangian-Eulerian)	
• Coupled Eulerian-Langrangian (CEL)	• For flow problems of structural problems with extensive deformations

setting-up an FE model, these steps are essentially the same. A typical sequence of steps in an FE simulation is illustrated and briefly described in Fig. 2.8. The simulation steps cover the setting-up of an FE model for the solution phase, submission of the job in the solution phase and visualizing the results as part of the analysis phase. Accurate data and information gathered about the problem during the mathematical modeling phase is input into the FE model through *GUI* of the FEA software used.

In the solution phase, the system of linear equations for the problem are simultaneously assembled and solved for the unknown degrees of freedom. These FE equations should have been coded in the commercial FEA software. However, users have some control over the type of analysis, the choice of solution methods, the rate of convergence and the associated level of numerical accuracy. A selected list of these choices and their brief descriptions are summarized in Table 2.2. Detailed descriptions should be available in users' manuals of the FEA software.

The FE analysis phase deals with post-processing of numerical results into meaningful graphical presentation and visualization. A strong background in physics, the mechanics of materials and materials science is required to extract meaningful information through interpretation of the calculated outcomes. Previous related experience should guide engineers in exercising good judgment and lots of common sense when extracting FE-predicted results for solutions of the problem.

Table 2.3 Checklist for "correct" FE results upon completion of FE solution phase

Obvious CHECKS for "correct" FE calculations
• Check imposed boundary conditions
• Check for consistency of deformed shape
• Check for appropriate order of the displacement magnitude
• Check for symmetry of the stress field
• Check for run-out of input material data/yield

Some obvious checks that can be performed immediately upon successful completion of the FE simulation job are listed in Table 2.3. Additional discussion on validating the FE model is discussed in Chap. 3 with emphasis on the solder joint reliability assessment.

References

1. Abaqus: Abaqus/CAE Users' Manual, Dassault Systemes Simulia Corp., RI (2012)
2. Anand, L.: Constitutive equations for hot working of metals. Int. J. Plast. **1**, 213–231 (1985)
3. Bath, K.J.: Finite element procedures. Prentice Hall, Englewood Cliffs (1996)
4. Huebner, K.H., Thornton, E.A., Byrom, T.G.: The finite element method for engineers. Wiley, New York (1982)
5. Hutton, D.V.: Fundamentals of finite element analysis. McGraw Hill, New York (2004)
6. Johnson, G.R., Cook, W.H.: Facture characteristics of three metals subjected to various strains, strain rates, temperatures and pressures. Eng. Fract. Mech. **21**(1), 31–48 (1985)
7. Reddy, J.N.: An introduction to the finite element method. McGraw Hill, New York (2006)
8. Tamin, M.N., Nor, F.M., Loh, W.K.: Hybrid experimental-computational approach for solder/IMC interface shear strength determination in solder joints. IEEE Trans Compon Packag Technol **33**(3), 614–620 (2010)

Chapter 3
Essentials for Finite Element Simulation

3.1 Introduction

Finite Element (FE) simulation solves the mathematical model of the physical problem. The following sections discuss the essential elements for the FE simulation process with respect to an illustrative problem in the assessment of solder joint reliability. The problem considers a surface mount microelectronic test assembly with a flip-chip package mounted on a Printed Circuit Board (PCB) using an array of solder joints. Figure 3.1a illustrates a typical cross-section of a microelectronic test assembly consisting of a silicon chip or die, an array of solder bumps with underfill, a substrate layer and an array of solder joints forming an interconnection to the PCB. The test assembly undergoes a solder reflow process and is subsequently subjected to reliability temperature cycles as illustrated in Fig. 3.1b. In this example, the reliability of the test assembly is assumed to be dictated by the reliability of these solder joints, particularly in the absence of the underfill. The objective of the FE simulation is then to establish the evolution of internal states of displacement, strain and stress in the critical solder joint of the array during solder reflow cooling, and subsequent temperature cycles. The evolutionary characteristics of inelastic strain on the critical solder joint is employed in assessing the reliability of the solder joints, thus the microelectronic test assembly.

The development of the mathematical model for the problem considers several simplifications of and assumptions on the various aspects of the problem. These include aspects of model geometry, an applied temperature profile and material constitutive laws. These simplifications and assumptions lead to an FE model with excellent computational efficiency, without sacrificing accuracy or the validity of the predicted physical phenomenon being studied. The essential requirements for the simulation process are deliberated in detail in the following sections. The mathematical modeling concept discussed in this chapter is repeatedly illustrated throughout the book for a range of different applications.

M. N. Tamin and N. M. Shaffiar, *Solder Joint Reliability Assessment*,
Advanced Structured Materials 37, DOI: 10.1007/978-3-319-00092-3_3,

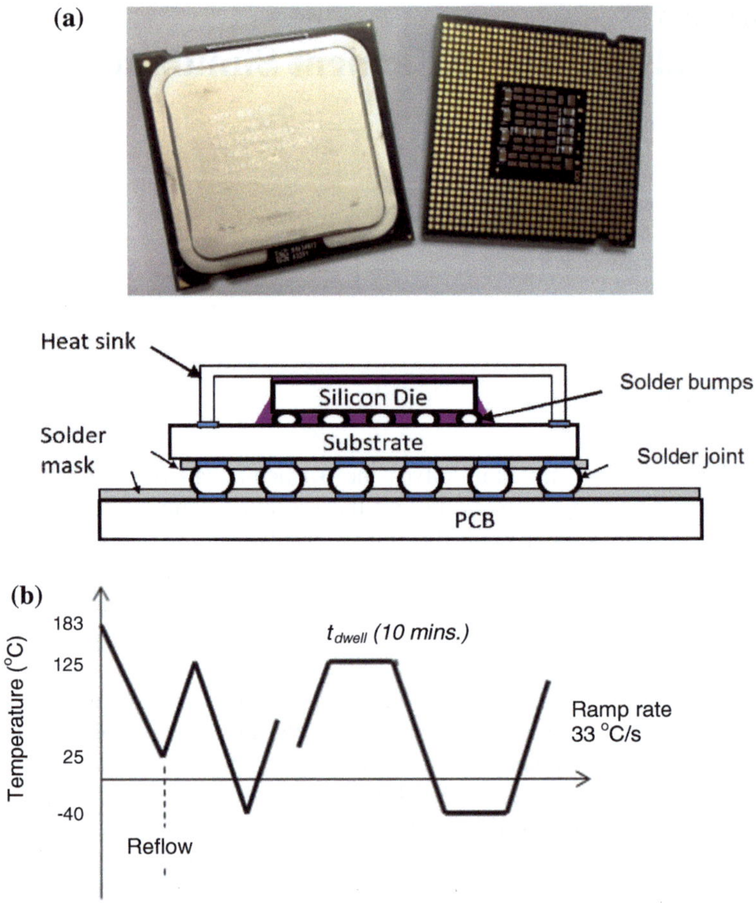

Fig. 3.1 **a** A cross section of a typical microelectronic test assembly with a flip chip package and BGA solder joints, and **b** temperature history for solder reflow and reliability temperature cycles. Values are typical for Sn-40Pb solder joints

3.2 Model Geometry

An accurate representation of the microelectronic test assembly, both in terms of the representative components and dimensions in the FE model, is required to ensure the accuracy of the predicted deformation. Dimensions of the various components are usually available from the design data of the package and assembly, and often can be measured with accuracy. In this example, the geometry of the assembly is assumed to inherit two planes of symmetry, namely plane *x*-*y* and plane *y*-*z*, thus only one quarter of the test assembly is sufficient to be modeled, as shown in Fig. 3.2a. The copper traces are also assumed to be symmetrically distributed with respect to these planes. The quarter FE model should also possess

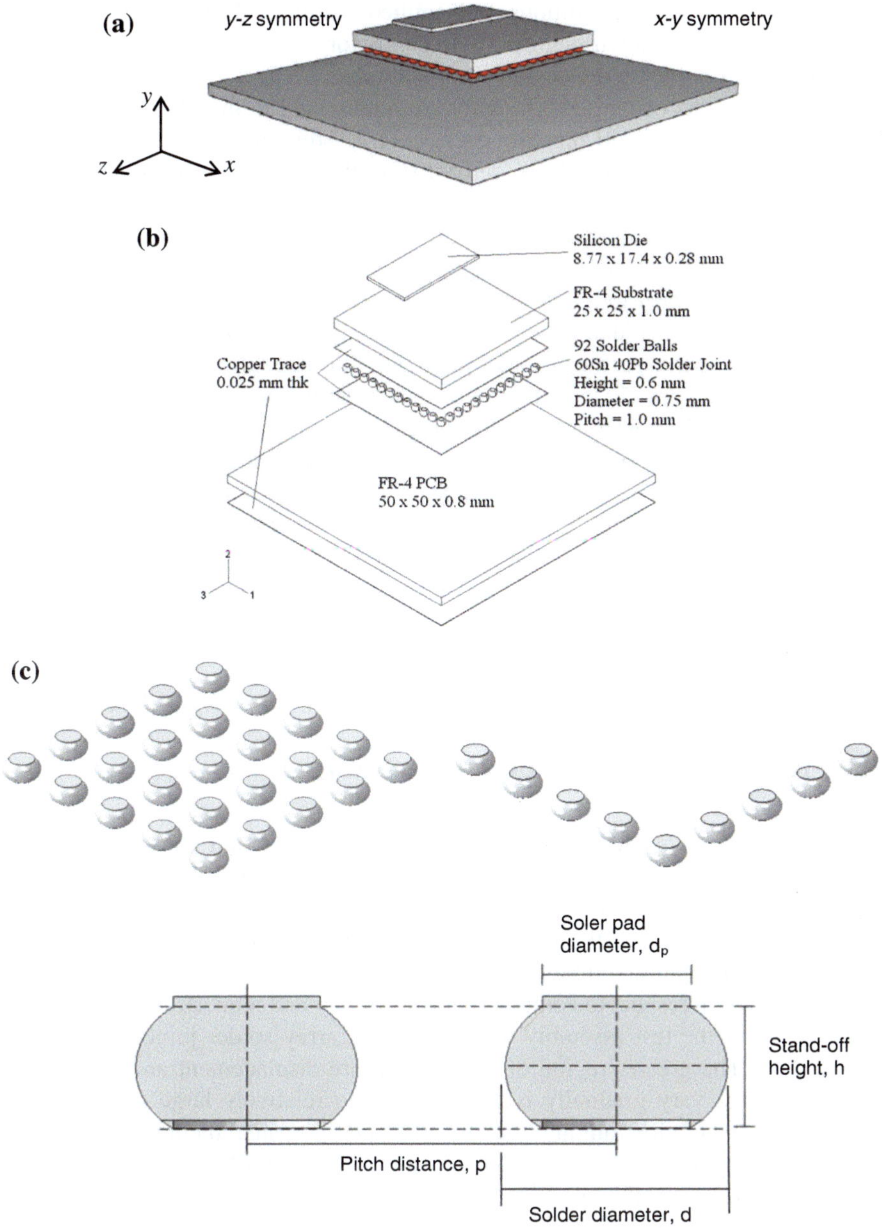

Fig. 3.2 **a** Symmetry planes (plane *x-y* and plane *y-z*) of the quarter model test assembly, **b** an exploded view of the quarter model, and **c** area array and peripheral array of solder joints (partial solder joints in the arrays are shown)

symmetry in boundary conditions and applied loadings. The quarter FE model reduces the total unknown degrees of freedom for the problem compared to the full model, thus reducing the cost of computations.

Key components of the test assembly are represented in the FE model, as illustrated in Fig. 3.2b. The stiff flip-chip package (consisting of the Si chip, the array of solder bumps and the underfill), as derived from the stiffness of the silicon chip, is modeled by an equivalent Si-die. Details of the solder bumps connecting the silicon die to the substrate and the underfill are excluded, as their contribution to the overall displacement of the test assembly is negligible. Copper traces found on the substrate and PCB are simplified to a uniform foil of thickness, 0.025 mm. Examples of two arrays of solder joints found in surface mount assemblies, namely an area array and peripheral array with one row of solder joints, are illustrated in Fig. 3.2c. Geometrical parameters relevant for FE modeling of the solder array are pitch distance, the maximum diameter of the solder and standoff height. In the illustrative example, a total of 92 near-eutectic Sn-40Pb solder joints are arranged in a peripheral array with a pitch distance of 1.0 mm. The maximum diameter of each solder is 0.75 mm, while the stand-off height is 0.6 mm. Solder mask-defined geometry at both devices (Si-die) and board side (PCB) of the solder joint is considered in the model. A thin pad of 0.02 mm-thick Cu_6Sn_5 intermetallic (IMC) phase is assumed to have formed and embedded between the solder joints and Cu traces at both device and board sides of the interconnections. Dimensions of other components are as shown in Fig. 3.2c.

3.3 Finite Element Mesh and Mesh Study

The size of elements and the degree of interpolation function chosen for the elements affect the accuracy of FE calculations of field variables, particularly in regions where the gradient of the variables is high. While fine-sized element mesh could ensure accuracy of the FE results, the large number of associated elements renders the computational time long and expensive. In this respect, FE mesh should be optimized with respect to computational efficiency and numerical accuracy.

FE mesh of the test assembly with peripheral array solder joints is illustrated in Fig. 3.3a. The greater region of the PCB where displacement and stress fields are expected to vary gradually is discretized into relatively large size elements. Mechanical interaction among the different components, including the solder joints, substrate, IMC layer, copper traces and PCB, could lead to complex stress fields, thus finer size elements are prescribed for these components in the vicinity of the solder joints. In addition, four layers of elements are prescribed through the thickness of PCB and substrate in order to yield an adequate accuracy of the displacement field (warpage) of the board throughout the analysis. The predicted warpage of the substrate greatly affects the deformation of the adjacent array of solder joints, thus influences the resulting strains and stresses in the solder.

Deformation of the most critical solder in the array with respect to failure is central to this problem. The FE predicted distribution and evolution of inelastic

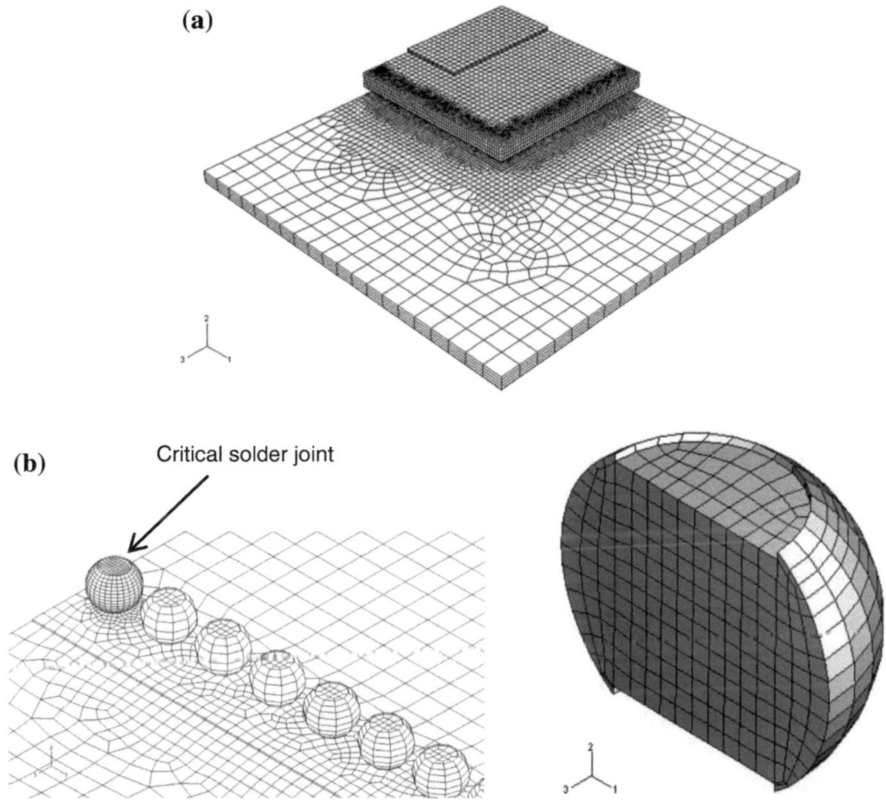

Fig. 3.3 **a** Finite element mesh of the test assembly illustrating the various mesh densities, and **b** refined mesh of the critical solder joint

strains and stresses in this critical solder joint will be examined rigorously with respect to the reliability of the test assembly. Thus, the critical solder joint should be discretized into elements with optimized mesh density. However, the location of the critical solder joint is often unknown a priori. In this case, an initial FE simulation of the test assembly with relatively coarse element mesh for all solder joints subjected to the reflow cooling process is performed. Results indicate that the solder joint located on the symmetry plane parallel with the longer side of the Si-die experienced the severest strain. Thus, the FE mesh for this solder joint is further refined as shown in Fig. 3.3b, for subsequent analysis.

As copper traces and Cu_6Sn_5 intermetallic pads are very thin, only one element throughout the thickness of each component is modeled. Linear incompatible mode elements are used to accurately reproduce the flexing behavior of these thin layers. The relatively stiff Si-package is also modeled with a layer of elements. In addition, an element layer of 0.05 mm-thick is prescribed in the critical solder joint adjacent to the solder/IMC interface. It follows from the observation that the fatigue crack path closely follows the pad-solder interface [5, 19]. These elements

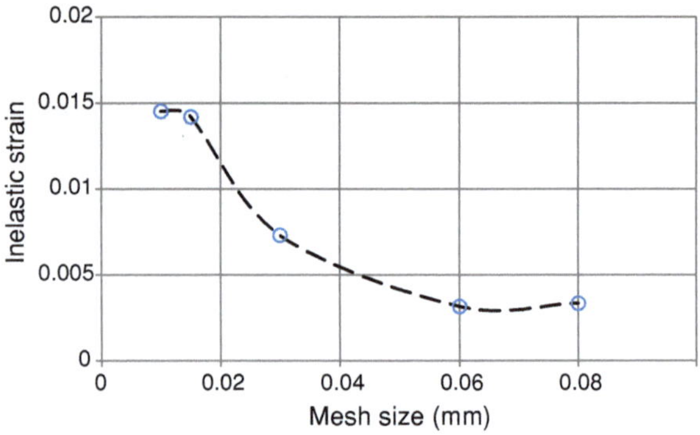

Fig. 3.4 Outcome of FE mesh convergence study for the critical solder joint in BGA assembly. Equivalent inelastic strain at the critical point of the solder joint is monitored

are used for averaging the plastic strain energy density to be employed in fatigue life calculations of the solder joint.

Continuum 3D elements, described as 8-node hexahedral solid or brick elements, are defined for the FE model. In this element, a linear variation of displacements in terms of nodal values is prescribed along each element edge containing only two nodes. Although a 20-node brick element is available in commercial FE analysis software, the computational cost involved is relatively higher due to the additional unknown degrees of freedom associated with a larger number of nodes and integration points. To this end, a 3D element with reduced integration points is also available as an option to shorten the analysis time. The total number of elements for the quarter model of the test assembly, as illustrated in this example, is 45,899.

A mesh convergence study refers to the systematic preliminary FE simulations performed on a given FE model to determine the largest acceptable size of the element that would yield mesh-independent results. The size of elements in the critical solder joint is systematically reduced for each subsequent simulation, while a selected monitoring variable, such as equivalent inelastic strain or von Mises stress, is recorded. Convergence of element mesh is achieved when the difference in values of the monitoring variable between successive simulations is acceptably small. An example of the outcome of the mesh-convergence study is shown in Fig. 3.4. The study establishes that the maximum size of an element with an edge length of 0.015 mm would give mesh-independent results.

3.4 Boundary Conditions

Boundary conditions are a description of known spatial positions of a point, edge or plane of the FE model geometry throughout the simulation. The conditions are often prescribed at the boundary of the model, and expressed in terms of known displacements and rotations. The symmetry boundary conditions are prescribed for the quarter model of the test assembly, as previously illustrated in Fig. 3.2a. The symmetry x-y plane should be constrained from any displacement along the z-axis direction because the plane is shared with the adjacent quarter test assembly. In addition, rotations of the x-y plane of the model with respect to the x- and y-axis are restricted for the same reason. These known linear and rotational displacements are expressed by the following boundary conditions for the x-y plane of the model:

$$U_z = U_{Rx} = U_{Ry} = 0 \tag{3.1}$$

Similarly, boundary conditions for the symmetry y–z plane of the model are prescribed as:

$$U_x = U_{Ry} = U_{Rz} = 0 \tag{3.2}$$

An arbitrary point on the model selected at the origin of the coordinate axes is also constrained from vertical displacement, $U_y = 0$, to exclude rigid body motion of the test assembly during analysis.

Since the FE simulation of the illustrated example will not consider separation between any components of the model, all surfaces initially in contact are assumed to be perfectly bonded throughout the reflow cooling and temperature cycling of the assembly. Different assumptions employed when simulating the damage and fracture conditions of the interface are covered in Chaps. 8 and 9.

3.5 Load and Load Cycles

Microelectronic packages and assemblies with BGA solders are subjected to the solder reflow process to form the solder joints. The controlled process exposes the assembly to prescribed temperature cycles in a reflow oven. The solder reflow temperature profile for an assembly with eutectic or near eutectic Sn-40Pb solders is shown in Fig. 3.5a. The solder reflow process begins with a preheating stage during which the solder paste dries while its more volatile ingredients evaporate. In the next stage when a temperature level of 150 °C is reached, the flux in the solder paste is activated to clean the bonded surfaces, thus improving wetting. At the end of the flux activation stage, all materials in the assembly should have reached the same temperature level. The assembly then enters the solder reflow stage as the solder alloy starts to melt at 183 °C. A short hold time period of about 60 s at higher than the melting temperature is prescribed to ensure complete melting and wetting of the solder, and to prevent solder bridging and cold solder joints. Solder joints begin to form as the molten solder solidifies at a temperature below 183 °C during the solder reflow cooling stage. Thermal strains and stresses are induced in the different materials making up the assembly as cooling progresses due to mismatches in the coefficient of thermal expansions of these materials. This reflow cooling process is modeled in the FE simulation. The process-induced residual strains and stresses, particularly in solder joints, raise various reliability issues of

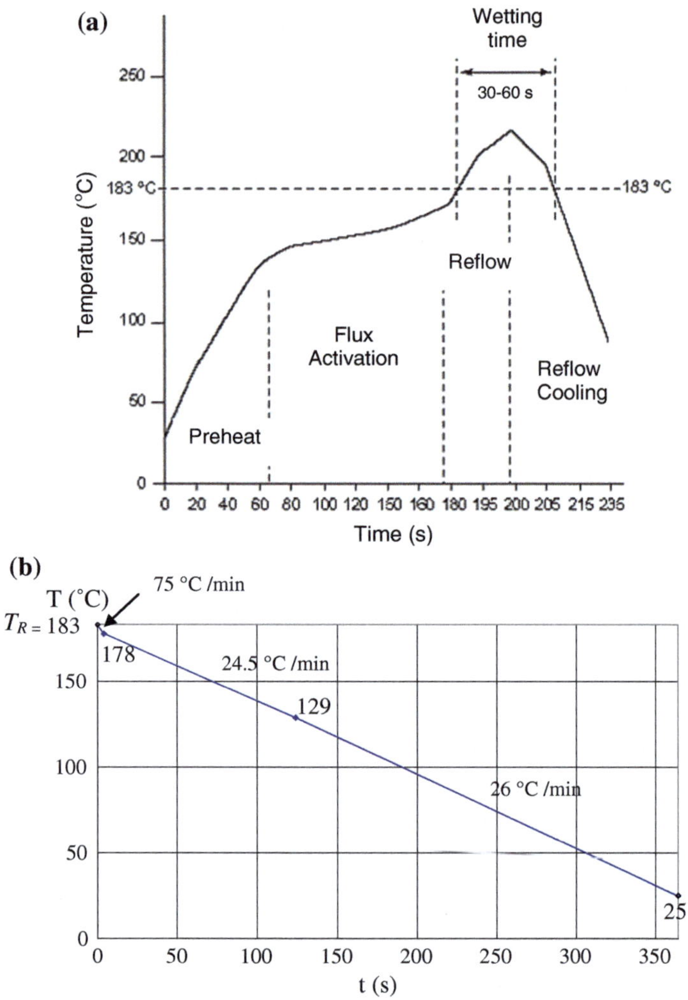

Fig. 3.5 **a** Reflow temperature history for BGA assembly with eutectic Sn–Pb solders, and **b** idealized reflow cooling temperature profile for FE simulation

the assembly. In assessing solder joint reliability, only the controlled cooling stage is considered in the solder reflow process simulation.

An ideal solder reflow cooling temperature history for FE simulation is shown in Fig. 3.5b. The Sn-40Pb solder joints and other materials in the assembly are assumed to be in their *stress-free* state at the solder reflow temperature, $T_R = 183$ °C. In addition, the assembly is assumed to be initially flat. Three different cooling rates are prescribed following the reflow cooling; 75, 24.5 and 26 °C/min. These rates closely reproduce the forced cooling effects of the reflow furnace on the assembly. The different cooling rates affect the mechanical response of the solder through the strain

Table 3.1 Sample of qualification test conditions for FCBGA packages [22]

Test environments	Conditions
Highly-accelerated stress test (HAST)	85 %RH[a]/85 °C
Autoclave	121 °C, 15 psig
Temperature cycles, air-to-air	−65/150 °C, or −55/125 °C −65/150 °C, or −40/125 °C
Thermal shock, liquid-to-liquid	−65/150 °C, or −55/125 °C, or −40 / 125 °C
High temperature operating life (HTOL)	125 °C, Op. voltage
HTOL[b]	140 °C, Op. voltage
HTOL[b]	155 °C, Op. voltage
Bake high temperature storage life (HSTL)	150 °C
Board level/solder joint reliability temperature cycles	−40/125 °C, or 0/100 °C
HAST	130 °C

[a] *RH* relative humidity
[b] One or more optional tests may be added to meet customer requirements

rate-dependent response of the solder alloy, discussed in the following sections. The solder reflow cooling process is completed when the room temperature reaches 25 °C, and solder joints and solder/IMC interfaces have formed.

A conventional life test of electronic products is conducted by monitoring the real time-to-failure of the products under normal use conditions. Since current products have a longer lifespan, an accelerated life test is more practical. An accelerated life test is based on either an accelerated usage or overstress of the product. Samples of qualification test conditions for FCBGA packages are summarized in Table 3.1 [22]. Reliability tests for electronic packages are often employed in Accelerated Temperature Cycles (ATC) and thermal shock cycles. In ATC tests, a package is placed in a temperature chamber and subjected to the prescribed temperature cycles until it fails. Failure of the solder joints is detected by monitoring electrical resistance of the solder joints [16]. Different load specifications for ATC tests on electronic packages and assemblies have been prescribed by manufacturers, testing bodies and related associations for package reliability requirements (e.g. Mil-STD-833, JEDEC No. 22-A104-B [9], IPC9701 [8] and NASA NHB5300.4) [6]. Sample ATC test conditions and requirements are listed in Table 3.2.

In the illustrated example, accelerated temperature cycles are applied to the test assembly with Sn-40Pb solder joints, as shown in Fig. 3.6. The temperature levels that vary from −40 to 125 °C correspond to the temperature range, $\Delta T = 165$ °C, and represent the design reliability temperature limit for the solder joints. The temperature ramp rate at 33 °C/min is significant as it dictates the strain rate- and temperature-dependent evolution of strains and stresses in the solder joints. Faster heating and cooling rates up to 370 °C/min have also been simulated to represent the thermal shock process on the assembly. In a reliability test, the thermal shock

Table 3.2 Sample of accelerated temperature cycling test conditions for reliability of electronic packages

Test condition	Mandated condition
Temperature cycles condition	85 %RH/85 °C
TC1	0 → +100 °C *preferred reference*
TC2	−25 → +100 °C
TC3	−40 → +125 °C
TC4	−55 → +125 °C
TC5	−55 → +100 °C
Test duration	Whichever condition occurs FIRST: 50 % (*preferred* 63.2 %) cumulative failure (preferred reference test duration) or number of thermal cycles
Requirement	
NTC-A	200 cycles
NTC-B	500 cycles
NTC-C	1,000 cycles (preferred for TC2, TC3, and TC4)
NTC-D	3,000 cycles
NTC-E	6,000 cycles (preferred reference TC1)
Low temperature dwell	10 min
Temperature, tolerance (preferred)	+0/−10 °C (+0/−5 °C) [+0/−18 °F (+0/−9 °F)]
High temperature dwell	10 min
Temperature, tolerance (preferred)	+10/−0 °C (+5/−0 °C) [+18/−0 °F (+9/−0 °F)]

Condition	Temperature cycles
A	−55/85 °C
B	−55/125 °C
C	−65/150 °C

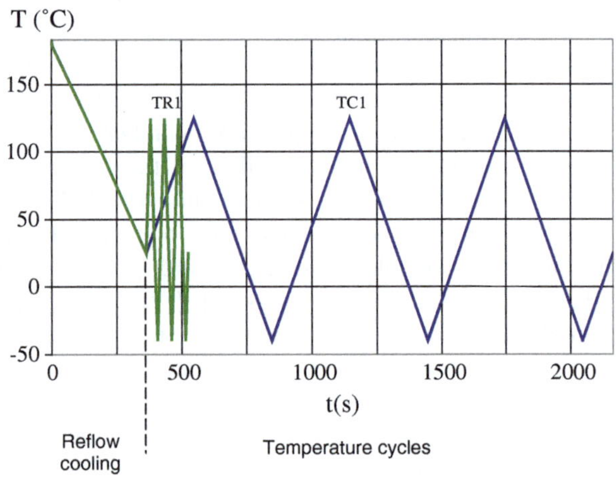

Fig. 3.6 Reliability temperature cycles (TC1) and thermal shock cycles (TR1) for assemblies with Sn-40Pb solder joints

process is performed by immersing the assembly in alternating hot and cold chemical baths. However, the actual temperature ramp rates are difficult to quantify. The accelerated temperature cycle and thermal shock cycle loading employed in this simulation is arbitrarily labeled as TC1 and TR1, respectively.

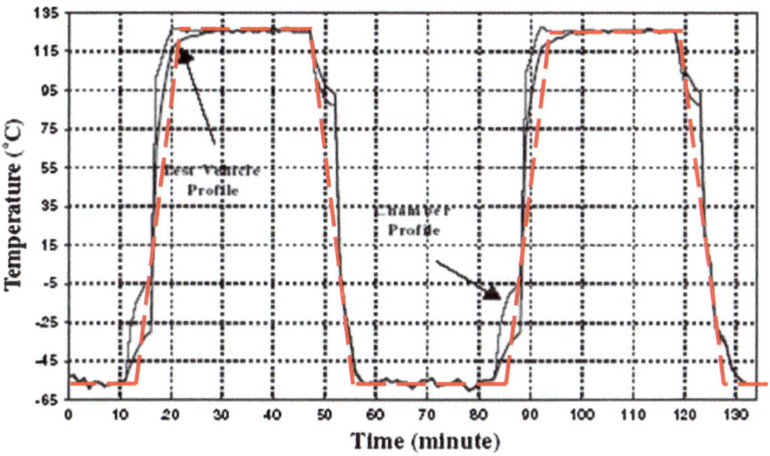

Fig. 3.7 Comparison of measured profiles [6] and idealized profile (*dashed line*) for temperature cycles with dwell periods for reliability test

Creep and creep-fatigue interaction effects on solder joint reliability can be examined by subjecting the assembly to temperature cycles with dwell periods at both hot and cold peak temperatures. Figure 3.7 shows a typical profile of the temperature cycle with a hold-time at high and low peak temperature levels [6]. An idealized temperature profile that can be used in FE simulation is shown for comparison as a dashed line. The prescribed temperature cycles vary from −55 to 125 °C, with a dwell period of 30 min each at peak temperature, for a total period of 68 min.

3.6 Material Models

Material models are mathematical equations describing the response of materials when subjected to a combination of stress, temperature and environment. These equations provide a unique relationship between stress and strain for every point in a material. The large temperature range during reliability temperature cycles induces inelastic thermal strains or plasticity while temperature cycles lead to fatigue of the solder joints in the test assembly. In addition, the high temperature level and long dwell time period results in the creep of the solder joints. Moreover, the fast temperature ramp rate is likely to induce viscoplasticity in the material. Material models for elastoplasticity, viscoplasticity, creep and fatigue are discussed in detail in Chap. 4. Selected material models for the various materials found in the illustrated problem and their rationale are presented in this section.

The stress–strain response of the near-eutectic Sn-40Pb solder employed in the illustrated problem is modeled based on the unified inelastic strain theory. The strain in the solder is assumed to be separable into an elastic component, ε_{ij}^{e}, and an inelastic component, ε_{ij}^{in}, such that:

$$\varepsilon_{ij} = \varepsilon_{ij}^{e} + \varepsilon_{ij}^{in} \tag{3.3}$$

The elastic strain is represented by Hooke's law in the form:

$$\varepsilon_{ij}^{e} = \frac{1+\nu}{E}\sigma_{ij} - \frac{\nu}{E}\delta_{ij}\sigma_{kk} \tag{3.4}$$

where E is the elastic modulus, ν is the Poisson's ratio and δ_{ij} is the Kronecker's delta, such that $\delta_{ij} = 1$, if $i = j$ and $\delta_{ij} = 0$, if $i \neq j$. Equation (3.4) is also known as the constitutive equation of the material. The hardening curves of the solder material are modeled by a set of constitutive equations, expressed in the time rate form of the inelastic strain, as [2]:

$$\dot{\varepsilon}_{in} = A\exp\left(-\frac{Q}{RT}\right)\left[\sinh\left(\xi\frac{\sigma}{s}\right)\right]^{1/m} \tag{3.5}$$

The term s in this equation captures the evolution of internal state variables in the solder throughout the loading, as expressed by:

$$\dot{s} = \left\{h_0\left|1-\frac{s}{s^*}\right|^{a}\cdot sign\left(\left|1-\frac{s}{s^*}\right|\right)\right\}\cdot\dot{\varepsilon}_{in} \tag{3.6}$$

with

$$s^* = \widehat{s}\left[\frac{\dot{\varepsilon}_P}{A}\exp\left(\frac{Q}{RT}\right)\right]^{n} \tag{3.7}$$

The flow equation described by Eqs. (3.5) through (3.7) accounts for the various inelastic strain components including creep and viscoplasticity in the solder. Thus, separate plasticity, creep and viscoplasticity model of the material is not required. Details of this unified inelastic strain constitutive model are described in Chap. 4. Values for model parameters are extracted from a series of true stress–strain curves of the material for different test temperature levels and strain rates in order to completely define the material model. Established parameter values for Sn-40Pb solders are listed in Table 3.3 [10]. Samples of true stress–strain curves for Sn-40Pb solders for different test temperature levels and strain rates, as predicted by the model, are compared with measured data in Fig. 3.8. Although the tension test data are shown for strain up to 2 pct., the material model is able to predict the curves beyond this strain level. Since the fracture model of the solder joint on the test assembly is not prescribed for the FE simulation, the calculated maximum strain on the solder joint should not reach the magnitude at fracture. In addition, the effective plastic strain at fracture in tension test specimens of a ductile solder material is dependent on the localized triaxial stress state at the necked section [15]. Material models for predicting damage and fracture of the solder and solder/IMC interface are discussed in Chap. 7. The different parts of the model that are initially in contact are assumed to be perfectly bonded throughout the temperature loading.

Table 3.3 Anand model parameter values for Sn-40Pb solder joint used in FE model [10]

Parameter	Description	Values
s_0	Initial value of state variable, s	5.643417
Q/R	Activation energy term	5,710.542 K
A	Pre-exponential factor	4,990.659 s^{-1}
ξ	Stress multiplier	2.547348
m	Strain rate sensitivity of stress	0.2305272
h_0	Hardening coefficient	60,235.88
$\hat{s}$	Coefficient for deformation resistance saturation value	31.21741 MPa
n	Strain rate sensitivity of the saturation value	0.1362387
a	Strain rate sensitivity of the hardening coefficient	1.578343

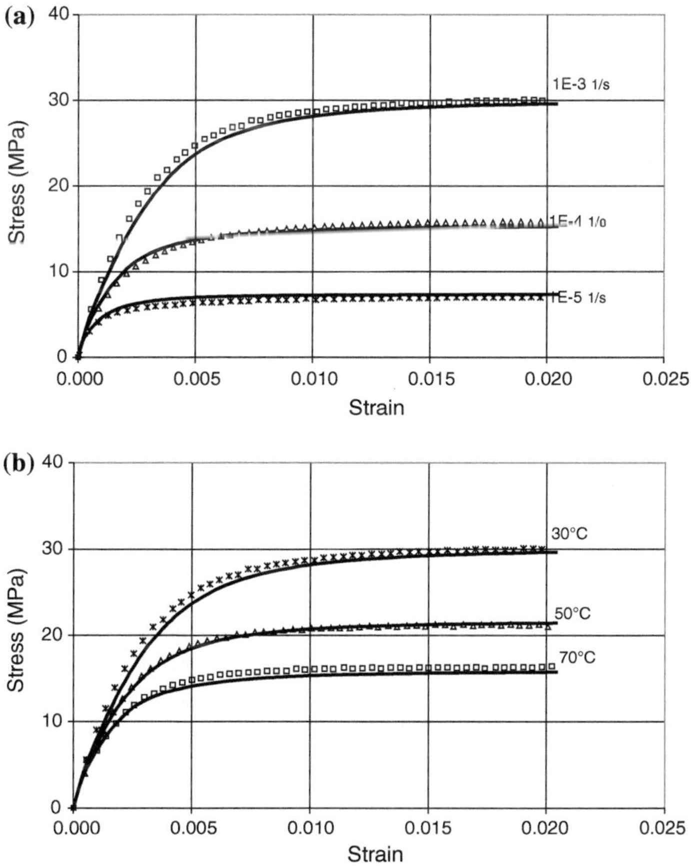

Fig. 3.8 Comparison of tensile test data and Anand model prediction for Sn-40Pb solder at **a** different strain rates, 30 °C, and **b** different test temperature levels, at strain rate of 10^{-3} s^{-1}. Test data is represented by the symbols [18]

Table 3.4 Temperature-dependent properties of different materials making-up the assembly

Material	E (GPa) T in Kelvin	ν	CTE, α (ppm/K) T in Kelvin
Sn-40Pb [18]	10	0.35	21
Silicon [20]	$132.46 - 0.00954\,T$	0.28	$2.113 + 0.00235\,T$
Copper [20]	$141.92 - 0.0442\,T$	0.35	$15.64 + 0.0041\,T$
Cu_6Sn_5 IMC [1]	85.6	0.31	16

Properties	−40 °C	30 °C	95 °C	125 °C	150 °C	270 °C
FR-4 (isotropic plane xy) [3]						
Ex (MPa)	24,252	22,400	20,680	19,300	17,920	16,000
Ey (MPa)	24,252	22,400	20,680	19,300	17,920	16,000
Ez (MPa)	2,031	1,600	1,200	1,000	600	450
Gxy (MPa)	662	630	600	500	450	441
Gxz (MPa)	210	199	189	167	142	139
Gyz (MPa)	210	199	189	167	142	139
Temperature-independent property						
ν_{xy}	0.02					
ν_{yz}	0.1425					
ν_{xz}	0.1425					
α_x (ppm/°C)	16					
α_y (ppm/°C)	16					
α_z (ppm/°C)	65					

Other materials in the illustrated FE model are assumed to behave elastically throughout the applied temperature cycles. The properties of the equivalent Si-package, copper traces and FR-4 substrate and PCB are also temperature-dependent. The FR-4 material is a Glass Fiber-Reinforced Polymer (GFRP) composite panel utilizing weaved glass fiber mat. The resulting orthotropic behavior of the FR-4 substrate and PCB is accounted for in the model. Properties of the various materials employed in the FE model of the test assembly are compiled in Table 3.4. Mechanical properties of these materials decrease while the coefficient of thermal expansion increases with increasing temperatures. These temperature-dependent variations can be represented using analytical equations or can be assumed to vary as a piecewise linear function for FE simulation. It is worth mentioning that the accuracy of the FE predicted response of solder joints and reliability of the test assembly is dictated by the prescribed properties and behavior of the various materials making up the assembly.

A fatigue life model is required for a reliability prediction of the test assembly. One such model, based on Coffin-Manson relation, is expressed as:

$$N_f = 1.2928(\Delta\gamma_{in})^{-1.96} \tag{3.8}$$

where N_f and $\Delta\gamma_{in}$ is the number of accumulated fatigue cycles or the fatigue life of the solder and the calculated inelastic shear strain range, respectively. The model postulates that shear strain range governs the fatigue life of the relatively ductile bulk solder joint. The coefficient and exponent of the strain-based model were determined for Sn-40Pb solder alloy at −50, 35 and 125 °C ([12, 22]). Other commonly employed fatigue models for solder joints are discussed in Chap. 7.

3.7 Aspects of Model Validation

Since the finite element method (FEM) essentially provides an approximate solution to the mathematical model of a given physical problem, an assessment of validity and accuracy of the predicted results is paramount. A validation process establishes the degree of correctness of the results, thus presenting the truth about the phenomenon being simulated. At best, selected FE predicted results are compared with relevant measured data from controlled experiments, the latter assumed to be correct. The validity of the FE model can also be established by comparing results of the FE simulation with an analytical solution for simple problems. The validated FE model is then used in subsequent analysis with realistic, often nonlinear, material behavior and complex loading.

Once the FE model is deemed valid, the accuracy of the calculated results provides the level of acceptance with respect to the intended use of the outcomes. Sources of computational inaccuracy are inherent with the solution approach employed in FEM, such as an implicit or explicit scheme, or a forward or backward difference marching scheme. A comprehensive discussion on computational efficiency and error analysis involved in FEM is covered elsewhere (for example [7, 14]). In performing parametric study, the absolute magnitude of a calculated variable, such as a shear stress, is not critical, as the values are often normalized by the shear strength of the material for analysis. In this case, one seeks for the trend in the data instead of focusing on the magnitude of the variable. On the contrary, accurate values of similar stress variables are critical when used in predicting the onset of a specific event, such as yielding or damage initiation at a specific material point.

It is argued here that the validation aspects of an FE model should begin before the simulation process. A material model for simulating solder joint behavior should be carefully examined for accurate reproduction of the non-linear stress–strain behavior of the solder alloy prior to utilization of the material model in FE simulation. Only an accurate and validated model could instill confidence in the predicted deformation behavior and the corresponding stresses and strains in the solder joint. Figure 3.8 compares the true stress–strain curves of Sn-40Pb solder with that predicted by the Anand model. The acceptably close comparison of the curves renders the material model valid for use in the FE simulation of solder joints experiencing different temperature levels and strain rates. Additional simulations of the uniaxial creep response of Sn-40Pb solder alloy are performed and the predicted creep curves are compared with measured data, as shown in Fig. 3.9. Reasonable comparisons of these curves provide validation of the constitutive model for use in the FE simulation of solder joints where extensive creep

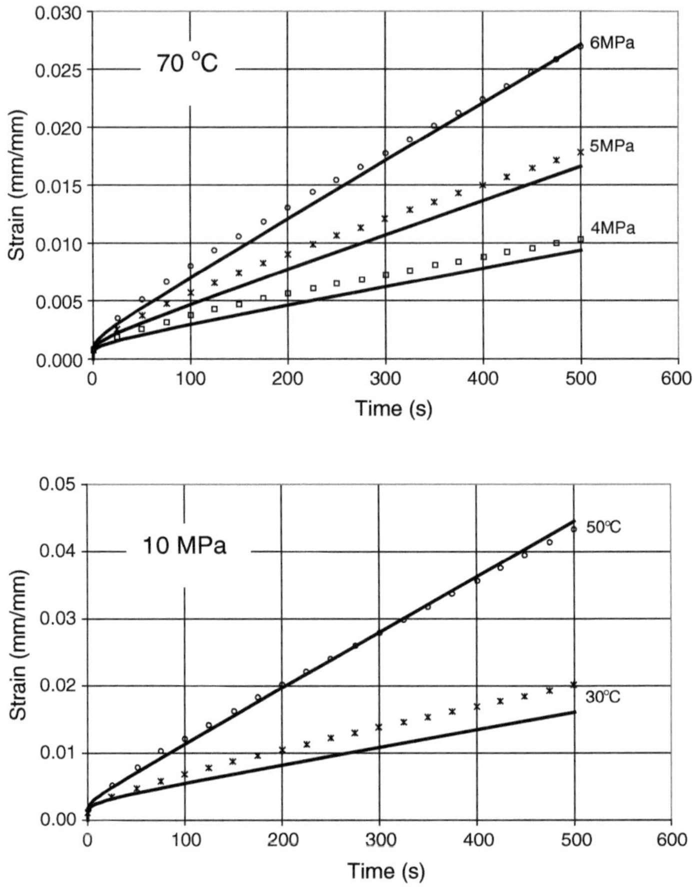

Fig. 3.9 A comparison of predicted creep curves by the Anand model with measured data points for Sn-40Pb solders at different applied stress and temperature levels. Test data from Sasaki et al. [18]

deformation is anticipated. The ability of a material model to predict the behavior within the range of loading conditions used for model validation purpose is often acceptable. However, extrapolation of the model for prediction of behavior beyond the validated range often requires good intuition and engineering judgment.

Upon completion of a simulation case, the results of the simulation should be checked for correct reproduction of the physical boundary conditions as represented by input values. The plot of the deformed shape of the FE model provides a visual inspection figure for assessing the boundary conditions. Figure 3.10 illustrates the deformed quarter model of the test assembly at 25 °C following the solder reflow cooling process. The fixed position of the node labeled *A* is reproduced, as indicated by zero displacement components. The symmetry *x*-*y* (1–2) plane and *y*-*z* (2–3) plane remain vertical throughout the loading.

A direct validation of the FE simulation results is assessed by comparing predicted and measured deformation or warpage of the test assembly following the

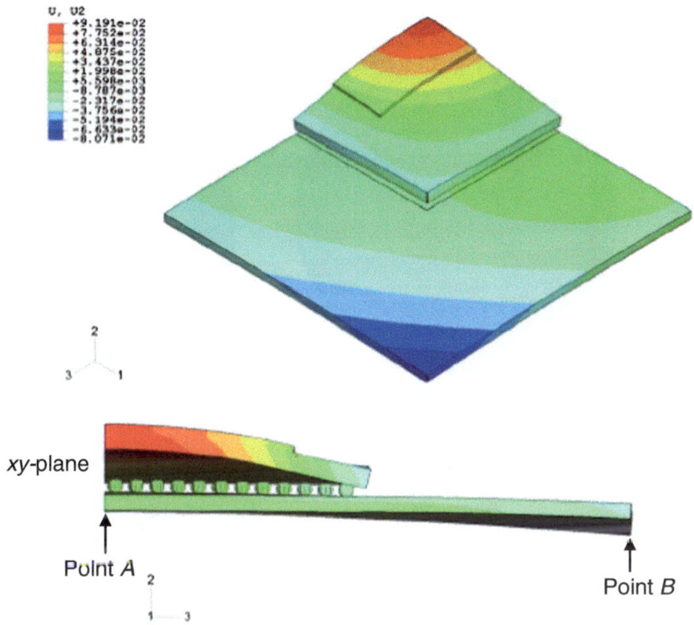

Fig. 3.10 The deformed quarter model of the test assembly at 25 °C following solder reflow cooling, illustrating the correct representation of the specified boundary conditions and warpage. Deformed scale is 15X

solder reflow process. Measurement of package warpage can be performed using the Moire method [4, 13]. A comparison of the shape of the warpage, either upwards resembling a bowl or a happy face or downwards showing a sad face and the maximum displacement, with measured value offers partial validation of the FE model in view of the complex geometry of the test assembly. Figure 3.10 illustrates the warpage of the test assembly with a sad face configuration.

Other aspects of FE model validation are argued through rationalizing the predicted response of the test assembly from an understanding of the mechanics of materials and limited experimental observations. Such information provides qualitative validation of the FE model.

3.8 Interpretation of Results

Results of the FE simulation are interpreted for information on displacements, strains and stresses, particularly in the critical solder joint that dictates the reliability of the test assembly. Such interpretation is meaningful only after validity of the FE model has been established, as discussed above. Results are presented as a quantitative description of meaningful variables with respect to the mechanics of the materials. Discussion of the results refers to reasoning and rationalizing with the purpose of providing physical understanding on the reported observation.

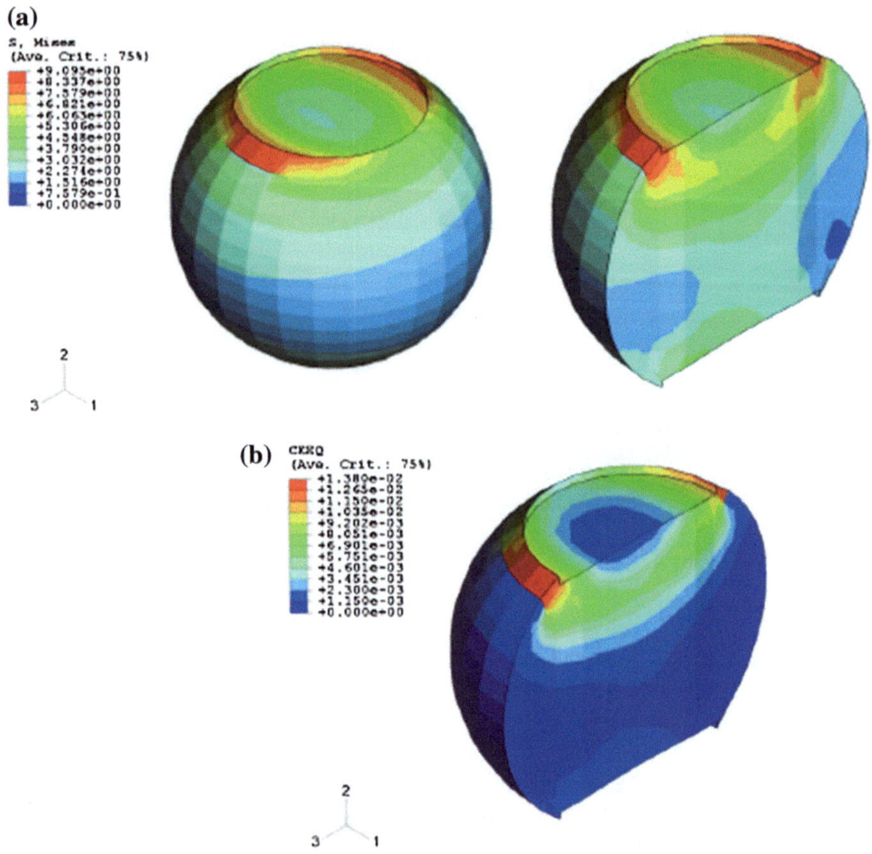

Fig. 3.11 **a** von Mises stress distribution in the critical Sn-40Pb solder joint at 25 °C following solder reflow cooling of the test assembly. The *right figure* shows the cut-out section of the solder, and **b** the corresponding equivalent inelastic strain in the critical solder joint

In assessing solder joint reliability, the evolution of critical variables including von Mises stress and inelastic strain throughout temperature cycling provides fatigue characteristics of the process. Spatial distribution of such variables in the critical solder indicates potential fatigue failure sites while the gradient of the variable reflects the degree of the localized failure event. The selected FE simulation results on the critical solder joint in the test assembly with peripheral array solders are presented in Fig. 3.11. An emphasis on interpretation of the results for various types of reliability-related information is highlighted.

Figure 3.11a shows the distribution of von Mises stress in the critical Sn-40Pb solder joint at 25 °C following the solder reflow process. The equivalent or von Mises stress is chosen in view of the ductile nature of the Sn-40Pb solder alloy. The stress scale is in MPa. The cut-out section of the solder joint shows internal stress distribution. A higher stress gradient is predicted for small regions in the solder joint

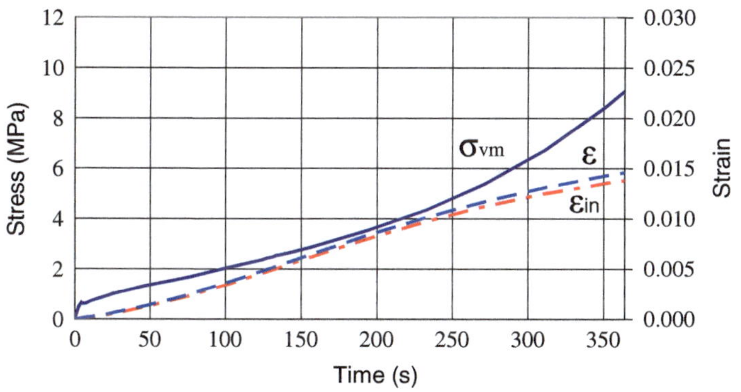

Fig. 3.12 Characteristic evolutions of von Mises stress, total and inelastic strain at the critical point in the solder joint throughout solder reflow process

near the edge of the solder/IMC interface on the device side of the test assembly. This induced stress is due to the effects of mismatches in the coefficient of thermal expansion and modulus of the metallic solder and adjacent Cu_6Sn_5 intermetallic layer. In this locality, yielding was predicted with the resulting maximum stress of 9.09 MPa and the corresponding accumulated inelastic strain of 1.38 pct., as illustrated in Fig. 3.11b. Crack initiation in this solder joint corresponds to the location of the maximum creep and plastic strain [17]. Fatigue crack is likely to initiate under subsequent reliability temperature cycles, as has been observed along the solder/IMC interface at the device side of similar assemblies with BGA packages [1, 11].

Characteristic evolution of von Mises stress and strains in the critical point of the solder joint is shown in Fig. 3.12. A nonlinear evolution of these variables are predicted following the nonlinear hardening model employed for the solder. Results indicate that inelastic strain initiated early following solder reflow cooling and continuously accumulated throughout the cooling process. This is attributed to the low yield strength of the solder relative to the evolving stress at high temperature levels. A close comparison of the total strain and inelastic strain curves indicates that this material point experiences extensive plasticity. However, the inelastic strain region is small and highly localized, while the bulk of the solder joint remain elastic, as indicated in Fig. 3.11b. Since residual internal states of stress and inelastic strain will be superimposed on the subsequent reliability temperature cycling of the test assembly, the influence on the reliability of the assembly should receive appropriate attention.

The accumulated inelastic strain can be viewed as a damage indicator to the fatigue process of the solder joint. Comparative evolutions of inelastic strain in the critical solder joint when subjected to different reliability temperature cycling profile (TC1) and thermal shock cycles (TR1) are shown in Fig. 3.13. The initial inelastic strain value was that induced during the solder reflow process. The nominal slope of the curve quantifying the nominal rate of inelastic strain accumulation per cycle could represent the rate of fatigue damage to catastrophic fracture. The larger magnitude of inelastic strain is achieved at the end of elapsed fatigue cycles with TC1

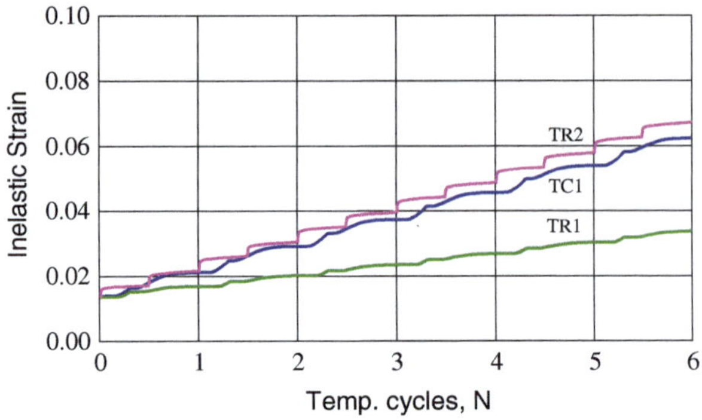

Fig. 3.13 Evolution of inelastic strain at the critical point in the Sn-40Pb solder joint for temperature cycles (TC1), thermal shock cycles (TR1) and modified thermal shock cycles with 10-min dwell time periods at peak temperature levels (TR2)

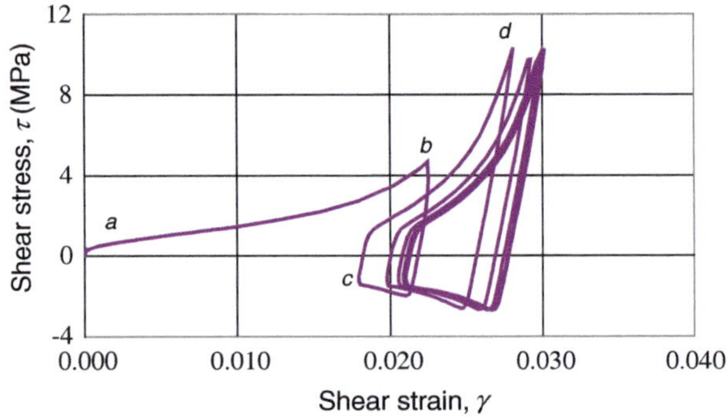

Fig. 3.14 Hysteresis loops for the shear stress–strain component in the critical solder joint during reliability temperature cycles (TC1)

primarily due to the creep effect over the relatively slow temperature ramp compared to that of the thermal shock cycles. A similar effect of creep contributing to the accumulated inelastic strain in the solder is demonstrated when a 10-min dwell time period at upper and lower peak temperature levels is included in the cycles (TR2).

The strain-based fatigue life model represented by Eq. (3.8) relies on the calculated inelastic shear strain range in the solder joint to establish solder joint reliability. The value of the shear strain range is taken from the results of the stabilized hysteresis loop calculated for the critical material points in the solder joint. Such hysteresis loops for the shear strain component resulting from reliability temperature cycles without dwell-time periods are shown in Fig. 3.14. Path *a–b* corresponds to reflow cooling from 183–25 °C, while the subsequent heating and cooling part of the first temperature cycle are represented by path *b–c* and *c–d* respectively. The inelastic shear strain range is defined as the range (maximum value–minimum value) of shear strain corresponding to zero shear stress. The magnitude of the shear strain range is then used in Eq. (3.8) for estimating the fatigue life of the solder joint corresponding to the applied reliability temperature profile. Availability of measured reliability data for identical test packages could be used to establish the validity of the FE model. The validated model could then be used with confidence to generate reliability test data for different geometry and reliability temperature cycles.

References

1. Amagai, M.: Chip scale package (CSP) solder joint reliability and modeling. Microelectron. Reliab. **39**(4), 463–477 (1999)
2. Anand, L.: Constitutive equations for hot working of metals. Int. J. Plast. **1**, 213–231 (1985)
3. Auersperg, J., Schubert, A., Vogel, D., Michel, B., Reichl, H.: Fracture and damage evaluation in chip scale packages and flip chip assemblies by FEA and Microdac. Symposium on Applications of Fracture Mechanics in Electronic Packaging, Dallas, AMD-vol 222/EEP-vol 20, pp. 133–138. ASME (ASM) (1997)
4. Cho, S.-M., Han, B., Joo, J.: Temperature dependent deformation analysis of ceramic ball grid array package assembly under accelerated thermal cycling condition. J. Electron. Packag. **126**, 41–47 (2004)
5. Dudek, R., Nylen, M., Schubert, A., Michel, B., Reichl, H.: An efficient approach to predict solder fatigue life and its application to SM and area array components. In: Proceedings of Electronic Components and Technology Conference, pp. 462–471 (1997)
6. Ghaffarian, R.: Qualification approaches and thermal cycle test results for CSP/BGA/FCBGA. Microelectron. Reliab. **43**, 695–706 (2003)
7. Huebner, K.H., Thornton, E.A., Byrom, T.G.: The Finite Element Method for Engineers. Wiley, New York (1982)
8. IPC 9701: Performance Test Methods and Qualification Requirements for Surface Mount Solder Attachments. Association Connecting Electronics Industries (IPC)
9. JESD22-A104B: Temperature Cycling. Electronic Industries Alliance/Joint Electron Device Engineering Council (EIA/JEDEC) (2000)
10. Koh, Y.K.: A unified constitutive models for solder materials. M. Eng. thesis, Universiti Teknologi Malaysia, Malaysia (2004)
11. Lee, T., Lee, J., Jung, I.: Finite element analysis for solder ball failures in chip scale package. Microelectron. Reliab. **38**, 1941–1947 (1998)
12. Manson, S.S.: Thermal stress and low cycle fatigue. McGraw-Hill, New York (1966)
13. Post, D., Han, B., Ifju, P.G.: High Sensitivity Moiré. Experimental Analysis for Mechanics and Materials Science and Technology. Springer, Berlin, Heidelberg (1994)
14. Reddy, J.N.: An Introduction to the Finite Element Method. McGraw Hill, New York (2006)
15. Rice, J.R., Tracey, D.M.: On the ductile enlargement of voids in triaxial stress fields. J. Mech. Phys. Solids **17**(3), 201–217 (1969)
16. Clech, J.P., Noctor, D.M., Manock, J.C., Lynott, G.W., Bader, F.E.: Surface mount assembly failure statistics and failure free time. In: Proceedings of 44th ECTC, pp. 487–497. IEEE (1994)
17. Sarihan, V.: Temperature dependent viscoplastic simulation of controlled collapsed solder joint under thermal cycling. ASME J. Electron. Packag. **115**, 16–21 (1999)
18. Sasaki, K., Ohguchi, K., Ishikawa, H.: Viscoplastic deformation of 40Pb/60Sn solder alloys experiment and constitutive modeling. ASME J. Electron. Packag. **123**, 379–387 (2001)
19. Shetty, S., Reinikainen, T.: Three and four point bend testing for electronic packages. ASME J. Electron. Packag. **125**, 556–561 (2004)
20. Shi, X.Q., Yang, Q.J., Wang, Z.P., Pang, H.L J., Zhou, W.: Reliability assessment of PBGA solder joints using the new creep constitutive relationship and modified energy-based life prediction model. In: Proceedings of 3rd Electronics Packaging Technology Conference (EPTC), pp. 398–405. IEEE (2000)
21. Solomon, H.D.: Fatigue of 60/40 solder. IEEE Trans. Compon. Hybrids, Manuf. Technol. **9**(4), 423–432 (1986)
22. Texas Instrument Inc.: Flip Chip Ball Grid Array Package Reference Guide (Unpublished) (2004)

Chapter 4
Mechanics of Solder Materials

4.1 Introduction

The mechanics of a material describes the response of a material to load. Such response is usually quantified in terms of displacement, strain and stress acting at every point in the material. The mechanical behavior of the material is represented using a stress–strain diagram. The diagram is obtained from tension test data on a sample of the material. Procedures for conducting a tension test on metallic materials are well documented in test standards such as ASTM-E8 [6] and ISO 6892 [9]. Some mechanical properties including elastic modulus, E, yield strength, S_Y and tensile strength, S_{UTS} of the material are determined from the stress–strain diagram. In addition, the diagram also indicates the plastic behavior of the material. Both properties and behavior are governed by its microstructure, which result from the processing of the material, such as casting, hot rolling, drawing and quenching. The processing–structure–property relationship that dictates the performance of a material is illustrated in Fig. 4.1. Equilibrium cooling of a cast rod, for example, can yield equiaxed grains in the metal rod. A subsequent drawing process of the metal rod into a wire results in elongated grains along the drawn direction. The resulting microstructure of the drawn wire exhibits greater strength and stiffness in the drawn direction. The temperature and strain rate during the tension test influence the resulting property values. Prolonged exposure to temperature and corrosive environments further degrades the intrinsic property values of the material. In addition, high homologous temperature, T/T_m where T_m is the melting point of the material, could initiate creep deformation, while load cycles lead to fatigue failure of the material. Consequently, it is necessary to accurately describe the mechanics of the material in predicting its response when subjected to single or combined loading conditions. This chapter describes material behavior under different loading types, namely elastic–plastic extensions, creep straining and fatigue loading. Examples of constitutive models that describe the stress–strain response of solder alloys are presented. A unified inelastic strain model that accounts for the combined effects of plasticity, viscoplasticity and creep is discussed. The performance

M. N. Tamin and N. M. Shaffiar, *Solder Joint Reliability Assessment*,
Advanced Structured Materials 37, DOI: 10.1007/978-3-319-00092-3_4,

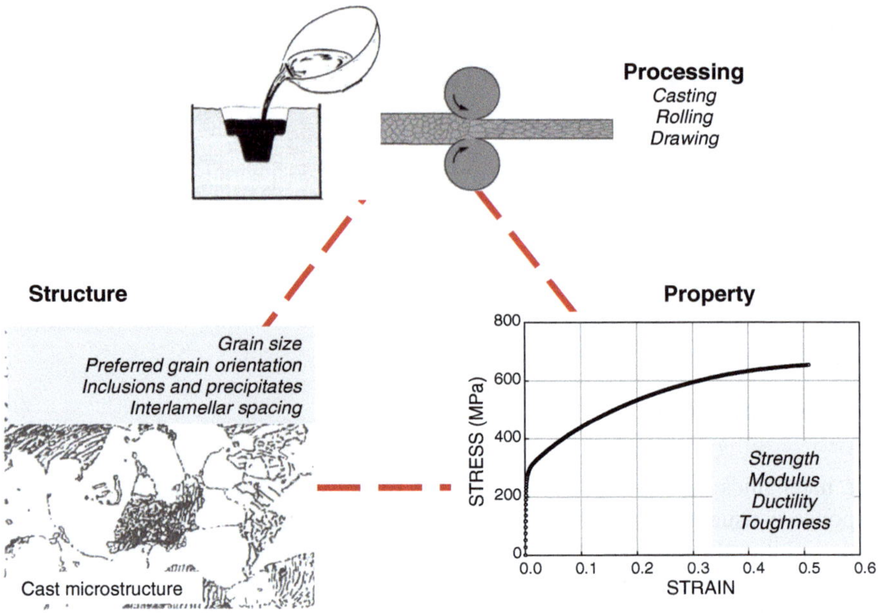

Fig. 4.1 Processing–structure–property relationship that defines the performance of a material

of the model in predicting the stress–strain behavior of lead-based and lead-free solder alloy over temperature and strain rate ranges of interest is illustrated. Aspects of the implementation of constitutive and fatigue models in FE simulations for the reliability assessment of the solder joints in BGA assembly are appropriately addressed.

4.1.1 Metallurgical Aspects of Solder Alloys

Solders are commonly divided into two categories, lead-based and lead-free solders. Lead-based solders, such as eutectic Sn–37Pb and near eutectic Sn–40Pb alloy, offer relatively low melting temperatures for the solder reflow process, while exhibiting excellent wetting and manufacturability, and good strength and fatigue characteristics for interconnections in electronic assembly. However, driven by legislative requirements for a greener environment and health-related concerns, an aggressive development of lead-free alternative soldering alloys are being pursued. Examples of lead-free solders and their respective melting points are listed in Table 4.1. The basic composition of these solders are Sn–Ag and Sn–Cu. Other compositions of Sn–Ag–Cu solders are also recommended [24, 37]. The associated higher melting temperature of lead-free solders raises the solder reflow process window over their leaded counterparts. This introduces reliability-related

Table 4.1 Examples of lead-based and lead-free solder alloys [23]

Chemical composition[a] (wt%)	Melting temperature (°C)
Sn–3.0Ag–0.5Cu–8.0In	195–199
Sn–3.5Ag–1.0B–4.0In	208–213
Sn–0.7Cu–0.5Ga–6.0In	210–215
Sn–3.1Ag-3.1Bi–0.5Cu	209–212
Sn–2.5Ag–0.8Cu–0.5Sb	217–219
Sn–3.5Ag–0.5Cu	217–218
Sn–4.0Ag–0.5Cu	221
Sn–3.5Ag	221
Sn–0.7Cu	227
Sn–37Pb (Eutectic)	183

[a]The balance is Sn

concerns for low-temperature materials of the substrate and high thermally-induced strains in the solder joints during solder reflow cooling and subsequent temperature cycling. In this respect, the addition of various doping elements, such as In, Ga, Sb and Bi is targeted for a lower melting point, improved microstructure stability and fatigue resistant properties.

Each solder alloy described by a specific chemical composition inherits a unique set of mechanical properties and behaviors. Experimental determination of these properties and behaviors for use in a reliability assessment of microelectronic assemblies with BGA packages is a daunting task. In this respect, a validated computational approach is indispensable for the simulation process and central to the theme of this book. For illustrative purpose, lead-based Sn–40Pb and lead-free Sn–4Ag–0.5Cu (SAC405) solders are used in this book to describe the various simulation requirements and demonstrate FE simulations of different processes in the assessment of solder joint reliability.

SAC405 solder alloy was introduced and published in 1965 [18] and is considered patent-free [29, 30, 46]. The partial top view of Sn–Ag–Cu ternary phase diagram is shown in Fig. 4.2. The diagram illustrates isotherms and the ternary eutectic point around 217 °C [40]. The composition of a SAC405 solder is indicated on the diagram with a melting point of 221 °C [19]. Equilibrium solidification of the alloy precipitates Ag_3Sn intermetallics. Although a SAC405 solder has a nominal chemical composition (in wt%) of 95.5 Sn, 4.0 Ag and 0.5 Cu, other elements may be present as impurities. Small variations in the chemical composition could lead to statistical variations in measured property values of the material.

The processing–structure–property relationship of a material calls for an appropriate examination of the resulting microstructure of solder joints following the solder reflow process. The typical microstructures of SAC405 solder on copper and *Electroless Nickel Immersion Gold* (ENIG) substrates in the vicinity of the solder/substrate interface are shown in Fig. 4.3a and b, respectively. Ag_3Sn intermetallic compound (IMC) with a plate-like structure grows into the bulk solder, while Cu_6Sn_5 intermetallic phase with a relatively short and rounded end forms on the continuous layer following solder reflow on a copper substrate. In reflowed

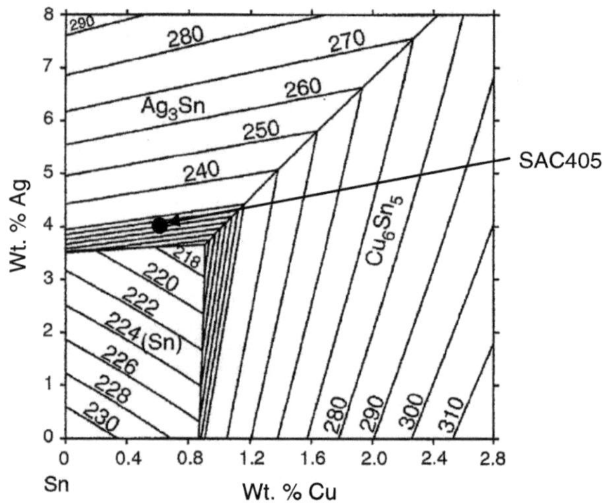

Fig. 4.2 Selected portion of the Sn–Ag–Cu ternary phase diagram indicating the composition of SAC405 solder alloy (http://www.metallurgy.nist.gov/phase/solder/solder.html)

SAC405 solder-on-ENIG substrate $(Cu, Ni)_6Sn_5$ IMC with a whisker-like shape grows at the solder/substrate interface along with finer-size Ni_3Sn particles. The rate of reflow cooling affects the initial microstructure, while subsequent thermal aging at combinations of exposure time duration and temperature levels further dictates the evolution of the microstructure in the solder joint. In addition, the intermetallic layer grows thicker during the thermal aging process. The growth rate of Cu_6Sn_5 IMC layer for the SAC405 solder on the ENIG and Cu substrate has been quantified in aging studies [41, 53]. Nucleation of new phases and coarsening of the Ag_3Sn IMC phase in the bulk solder joint has also been reported. Similar effects of temperature and the aging time duration on IMC nucleation and growth phenomena occur for BGA solder joints in microelectronic assemblies under reliability temperature cycles. Such complex structure of the solder/IMC interface region and the growth mechanics of IMC layer are often simplified in FE simulations of the solder joint response in BGA packages and assemblies for practical reliability assessment.

It is common to assume that a uniform Cu_6Sn_5 IMC layer has formed following the solder reflow process and does not grow thicker during subsequent temperature cycling in the mathematical model for FE simulation of the solder joint. In addition, the hard and brittle IMC phase is assumed to behave as a linear elastic material with temperature independent properties throughout the applied temperature loading. Furthermore, details of surface morphology of the IMC layer is not considered in the simulation due to the associated large number of local feature variations. The incorporation of such microscopic details, while increasing the computational power requirement, would not yield significant refinement in the results.

The microstructure of a reflowed solder joint undergoes grain coarsening during the shelf life of the material. This phenomenon is driven by internal stresses at the microstructure level. Subsequently, the degradation of tensile properties following room temperature aging of SAC405 solder alloy has been quantified [33]. In this respect, several constitutive models incorporating microstructure-coarsening

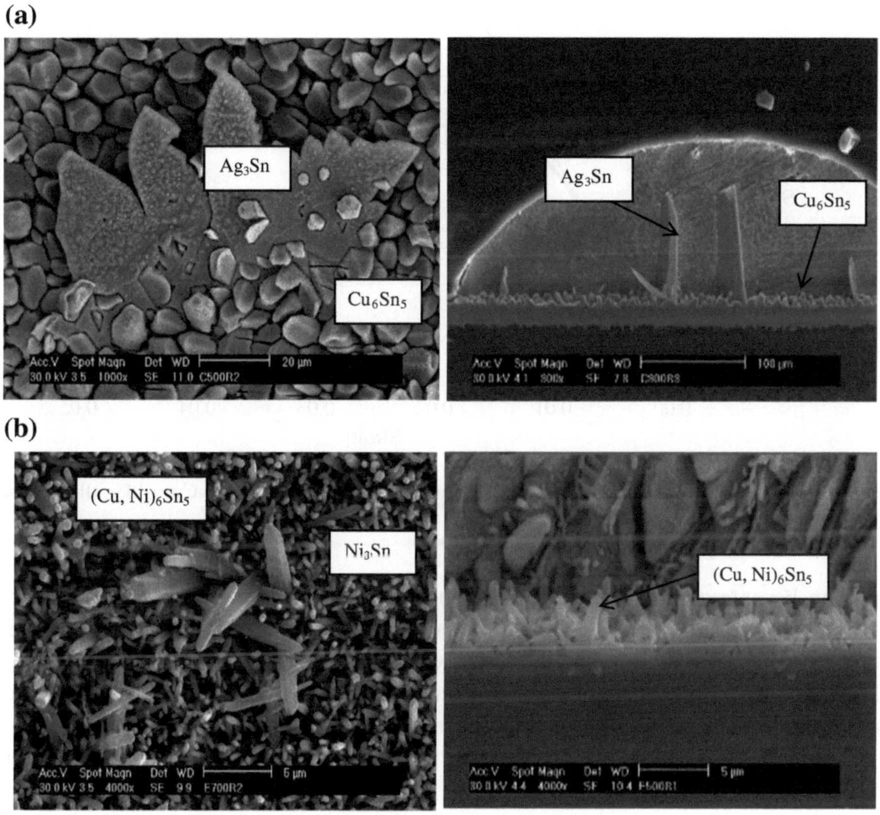

Fig. 4.3 Microstructure of reflowed SAC405 solder on (**a**) copper substrate and (**b**) ENIG substrate showing the different phases at the solder/substrate interface region. Plan view (*left*) and cross-section view (*right*)

effects have been introduced [15, 16, 20, 49, 50]. Such a model could be employed in establishing the time-dependent property-structure degradation behavior of the material. However, this phenomenon is not being modeled in the FE simulation examples discussed in this book.

4.2 Elastic Behavior and Yielding

The characteristic behavior of numerous metals is often represented by Hooke's law in the elastic range. An example of such linear elastic behavior is illustrated using the initial part of the stress–strain diagram for Sn–Ag–Cu solder, as shown in Fig. 4.4. Several mechanical properties of the material, such as elastic modulus, E represented by the slope of the initial linear part of the curve, and yield strength,

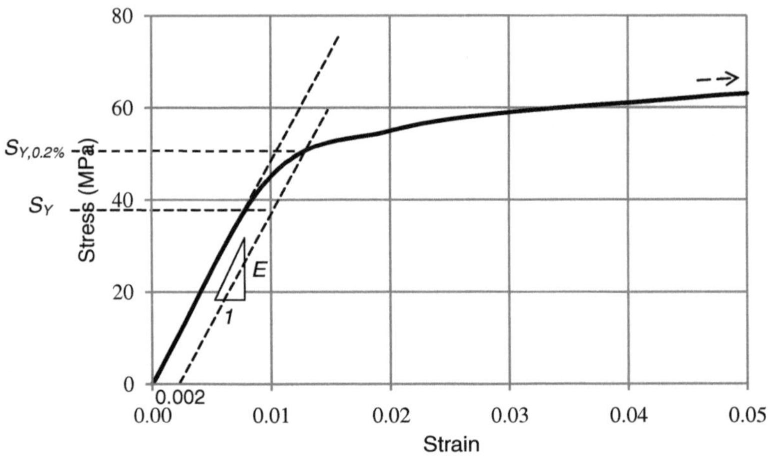

Fig. 4.4 The initial part of the stress-strain diagram for a typical Sn–Ag–Cu solder alloy showing initial elastic behavior

S_Y, can be extracted from the data, as indicated in the figure. It is noted that the common definition of 0.2 % offset yield strength, $S_{Y,0.2\%}$, is not appropriate in view of representing the elastic–plastic curve of the material for FE simulation. Therefore, yield strength of the solder alloy, S_Y, is defined as the stress level corresponding to the first deviation from linearity in the stress–strain curve. In addition, the solder alloy is assumed to be homogeneous and isotropic when considering the mechanics of the material.

4.2.1 Generalized Hooke's Law

Since the general stress and strain states in the solder joint are complex, 3D Cartesian stress and strain components are used to describe the internal states of the material. The generalized Hooke's law is an explicit relationship in relating stresses to strains at a material point under elastic condition. The constitutive equations are reproduced here, without any proof, for convenience:

$$\begin{Bmatrix} \sigma_x \\ \sigma_y \\ \sigma_z \\ \tau_{xy} \\ \tau_{xz} \\ \tau_{yz} \end{Bmatrix} = \frac{E}{(1+v)(1-2v)} \begin{bmatrix} 1-v & v & v & 0 & 0 & 0 \\ v & 1-v & v & 0 & 0 & 0 \\ v & v & 1-v & 0 & 0 & 0 \\ 0 & 0 & 0 & \frac{1-2v}{2} & 0 & 0 \\ 0 & 0 & 0 & 0 & \frac{1-2v}{2} & 0 \\ 0 & 0 & 0 & 0 & 0 & \frac{1-2v}{2} \end{bmatrix} \begin{Bmatrix} \varepsilon_x \\ \varepsilon_y \\ \varepsilon_z \\ \gamma_{xy} \\ \gamma_{xz} \\ \gamma_{yz} \end{Bmatrix} \tag{4.1a}$$

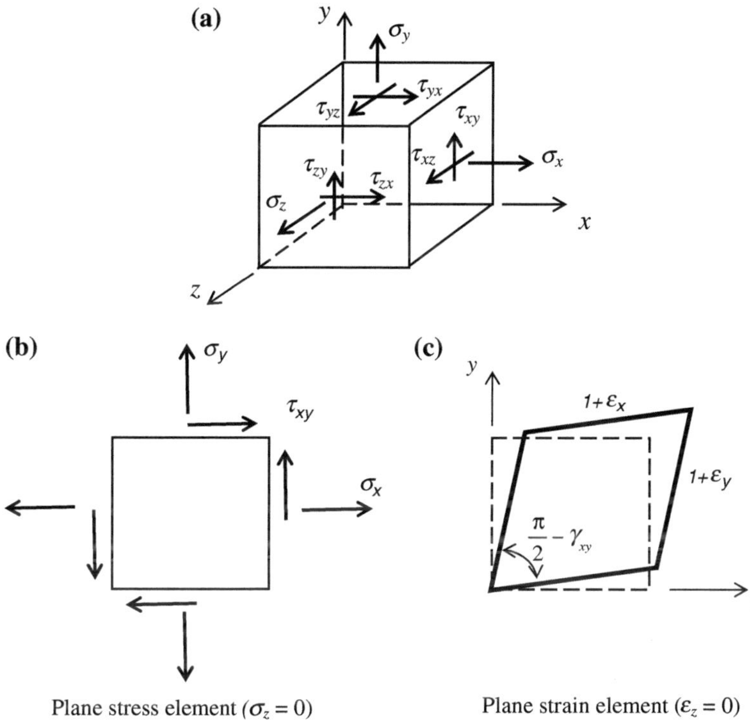

Fig. 4.5 Cartesian coordinate axes and definition of positive stress and strain components. **a** General 3D stress state, **b** plane stress element and **c** plane strain element representing the respective condition

or alternately expressed in compact form as:

$$\{\sigma\} = [C]\{\varepsilon\} \quad (4.1b)$$

where σ_x, σ_y, σ_z are normal stress and τ_{xy}, τ_{xz}, τ_{yz} are shear stress components. The terms ε_x, ε_y, ε_z are the corresponding normal strain and γ_{xy}, γ_{xz} and γ_{yz}, the corresponding shear strain components. The matrix, $[C]$ in Eq. (4.1b) is often called the elasticity matrix. The stressed element, as illustrated in Fig. 4.5a, defines the Cartesian coordinate axes for these stress components. Static equilibrium of the stress element requires that $\tau_{xy} = \tau_{yx}$, $\tau_{yz} = \tau_{zy}$ and $\tau_{xz} = \tau_{zx}$. The terms E and ν represent elastic modulus and Poisson's ratio, respectively. The latter defines the transverse contraction of a material element with respect to its elongation in the applied load direction. In the case where the shear modulus, G of the material is experimentally determined, each of the shear traction components can be calculated as:

$$\tau = G\gamma \quad (4.2)$$

Otherwise, the shear modulus can be estimated from the following equation:

$$G = \frac{E}{2(1+\nu)} \quad (4.3)$$

In limited cases where 3D geometry, loading and boundary conditions can be simplified for the deformation and stress analysis into either plane stress or plane strain condition, Eqs. (4.1a) and (4.1b) is reduced accordingly. In a plane stress condition where the stress on the designated z-plane diminishes, i.e. $\sigma_z = 0$, $\tau_{xz}(=\tau_{zx}) = 0$, $\tau_{yz}(=\tau_{zy}) = 0$, the constitutive behavior is given by:

$$\begin{Bmatrix} \sigma_x \\ \sigma_y \\ \tau_{xy} \end{Bmatrix} = \frac{E}{(1-\nu^2)} \begin{bmatrix} 1-\nu & \nu & 0 \\ \nu & 1-\nu & 0 \\ 0 & 0 & \frac{1-\nu}{2} \end{bmatrix} \begin{Bmatrix} \varepsilon_x \\ \varepsilon_y \\ \gamma_{xy} \end{Bmatrix} \tag{4.4}$$

The non-zero strain along the normal z-direction can be computed independently of Eq. (4.4) as:

$$\varepsilon_z = -\frac{\nu}{E}\left(\sigma_x + \sigma_y\right) \tag{4.5}$$

Similarly, a condition of plane strain exists when the strain component along the designated z-coordinated direction diminishes, i.e. $\varepsilon_z = 0$, $\gamma_{xz}(=\gamma_{zx}) = 0$, $\gamma_{yz}(=\gamma_{zy}) = 0$. The generalized Hooke's law is then reduced to:

$$\begin{Bmatrix} \sigma_x \\ \sigma_y \\ \tau_{xy} \end{Bmatrix} = \frac{E}{(1+\nu)(1-2\nu)} \begin{bmatrix} 1-\nu & \nu & 0 \\ \nu & 1-\nu & 0 \\ 0 & 0 & \frac{1-2\nu}{2} \end{bmatrix} \begin{Bmatrix} \varepsilon_x \\ \varepsilon_y \\ \gamma_{xy} \end{Bmatrix} \tag{4.6}$$

The non-zero stress component on the z-plane can then be evaluated independently of Eq. (4.6) as:

$$\sigma_z = \nu\left(\sigma_x + \sigma_y\right) \tag{4.7}$$

The state of plane stress and plane strain is represented by stress element and strain element, respectively, as illustrated in Fig. 4.5b.

The property values of elastic modulus, E, for the SAC405 solder are affected by both temperature and strain rates. The variations are shown in Fig. 4.6 for a range of test temperature levels between −40 and 180 °C and applied strain rates up to 0.0075 s^{-1}. The temperature-dependent property can also be represented by a best-fit line through the data using the least-squared method. The elastic modulus of the solder decreases with an increasing temperature, while the material displays a higher modulus at a higher straining rate.

4.2.2 Thermal Strains

Non-uniform temperature distribution affects the behavior of the linear elastic material through the contribution of thermal strains and the corresponding stresses. In an isotropic material, the induced strains due to thermal expansion

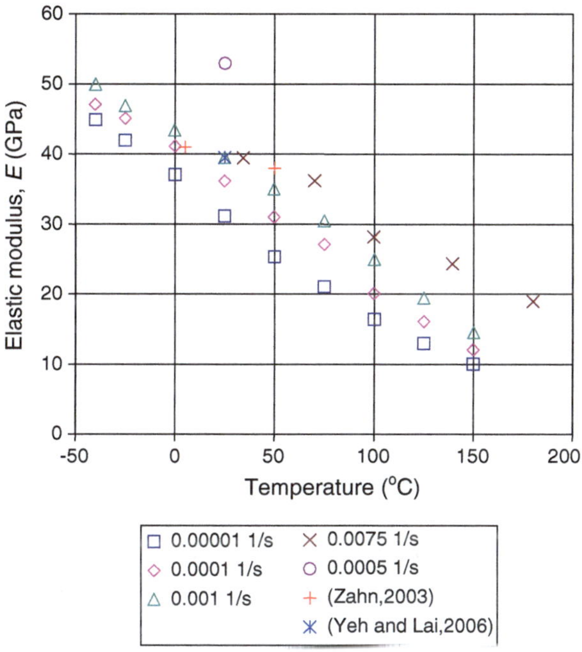

Fig. 4.6 Variation of elastic modulus values for SAC405 solder with temperature and strain rates [32, 52, 55, 64, 65]

are uniform in all directions, thus contribute only to normal strain components. Consequently, the thermal strain vector is expressed as:

$$\{\varepsilon\}_T = \left\{ \begin{array}{c} \varepsilon_x \\ \varepsilon_y \\ \varepsilon_z \\ \gamma_{xy} \\ \gamma_{xz} \\ \gamma_{yz} \end{array} \right\}_T = \left\{ \begin{array}{c} \alpha(T - T_o) - \alpha(T_i - T_o) \\ \alpha(T - T_o) - \alpha(T_i - T_o) \\ \alpha(T - T_o) - \alpha(T_i - T_o) \\ 0 \\ 0 \\ 0 \end{array} \right\} \tag{4.8}$$

where α is the Coefficient of Thermal Expansion (CTE) that defines the total thermal expansion of the material from a reference temperature, T_o. The reference temperature term is necessary to enforce the condition of no initial strain for cases in which the reference temperature differs from the initial temperature. The terms T and T_i are the current and initial temperatures, respectively. CTE values can also be temperature-dependent, as illustrated for the SAC405 solder alloy shown in Fig. 4.7 [39]. The stress–strain equation, accounting for the thermal effect, can then be expressed as:

$$\{\sigma\} = [C]\{\varepsilon\} + [C]\{\varepsilon\}_T \tag{4.9}$$

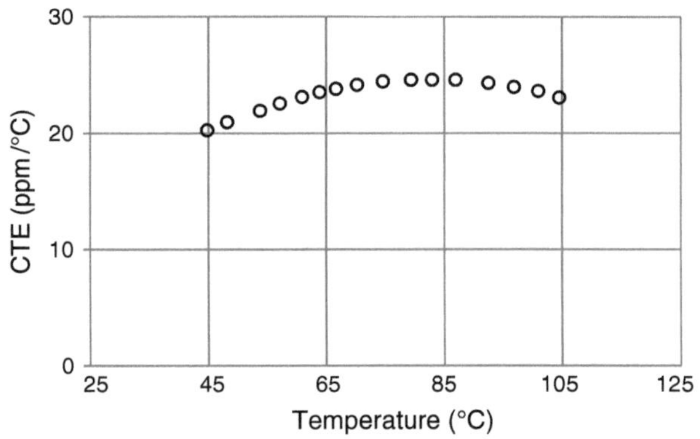

Fig. 4.7 Variation of the coefficient of thermal expansion values for SAC405 solder with temperature. Data extracted from Nguyen and Seungbae [39]

4.2.3 Yield Criteria

A yield criterion is required to indicate the stress level at which plastic flow commences in the material. The yield criterion, F can be written in general form as:

$$F(\{\sigma\},\kappa) = f(\{\sigma\}) - k(\kappa) = 0 \tag{4.10}$$

where $f(\{\sigma\})$ is some function of stress and $k(\kappa)$ is a material parameter, which could be a function of a hardening parameter, κ. It is worth emphasizing that a yield criterion should be independent of the orientation of the coordinate system.

During the tension test of a ductile material, such as a solder alloy, yielding is caused by the slipping of the randomly ordered crystals making up the material due to shear stress. Thus, at an initial yield, the maximum shear stress in the uniaxially loaded specimen would have reached the shear yield strength level of the material. Based on a stress transformation (or Mohr's circle) for the uniaxial stressed element, the maximum shear stress at yield $(\tau_{max})_Y$ is equivalent to one-half the yield strength value and occurs on a plane oriented at 45° to the plane with applied normal stress, i.e.:

$$(\tau_{max})_Y = \frac{S_Y}{2} \tag{4.11a}$$

This forms the basis for the Maximum-shear-stress (Tresca) theory. Equation (4.10) can be stated for the onset of yielding to occur when the absolute maximum shear stress at the critical point in a material equals the shear yield strength, as observed in a tension test. The general state of stresses, as illustrated in Fig. 4.5, can also be expressed in terms of principal stress components, σ_1, σ_2 and σ_3. If the principal stresses are numerically arranged along the stress axis such that $\sigma_3(=\sigma_{min}) < \sigma_2 < \sigma_1(=\sigma_{max})$, then the Tresca yield criterion can be written as:

$$\sigma_{max} - \sigma_{min} = S_Y \tag{4.11b}$$

The LHS term of Eq. (4.11b) is called Tresca stress, as used in FE analysis, and represents the stress function in Eq. (4.10).

An alternate way of defining the onset of yielding in a ductile material is to consider the strain energy density stored internally in a loaded material. The general state of stresses in an elastic material can be decomposed into hydrostatic (uniform) stress components causing a change in volume and dilatational stress components resulting in distortion of the material element. The former does not cause yield in the material as has been demonstrated experimentally [8]. Thus, the distortional strain energy density u_d can be expressed using principal stresses as:

$$u_d = \frac{1+v}{6E}\left[(\sigma_1-\sigma_2)^2+(\sigma_2-\sigma_3)^2+(\sigma_3-\sigma_1)^2\right] \tag{4.12}$$

During the tension test, the distortional strain energy density at initial yield $(u_d)_Y$ can be inferred from Eq. (4.12) with $\sigma_1 = S_Y$, $\sigma_2 = \sigma_3 = 0$. Thus:

$$(u_d)_Y = \frac{1+v}{3E}(S_Y)^2 \tag{4.13}$$

This becomes the basis for the Maximum-distortional-energy (von Mises) theory. Based on Eq. (4.10), the onset of yielding in a material subjected to complex stresses occurs when the maximum distortional strain energy density in the material equals the value at the yield for an identical material in a tension test, i.e. $u_d = (u_d)_Y$, or:

$$\left[(\sigma_1-\sigma_2)^2+(\sigma_2-\sigma_3)^2+(\sigma_3-\sigma_1)^2\right] = 2(S_Y)^2 \tag{4.14a}$$

Equation (4.14a) can be re-arranged in another form, commonly used in FE analysis when discussing yielding of a material:

$$\sqrt{\frac{1}{2}\left[(\sigma_1-\sigma_2)^2+(\sigma_2-\sigma_3)^2+(\sigma_3-\sigma_1)^2\right]} = S_Y \tag{4.14b}$$

The LHS term of Eq. (4.14b) is called von Mises stress and represents another stress function described in Eq. (4.10) for the yield criterion.

Yield surface is a graphical representation of the yield criterion in the principal stress space. Such geometrical representation of the Tresca and von Mises yield surfaces is illustrated in Fig. 4.8a as a projection of the yield surface onto the π-plane, described by $\sigma_1 + \sigma_2 + \sigma_3 = 0$. A more common representation of yield loci plotted in the principal plane stress space is shown in Fig. 4.8b. A calculated stress state located inside the locus represents an elastic condition, while that on or outside the locus of the initial yield has plastically deformed. While the von Mises equation better fits the experimental data than Tresca criterion for many metals, the latter is simpler to use in theoretical applications.

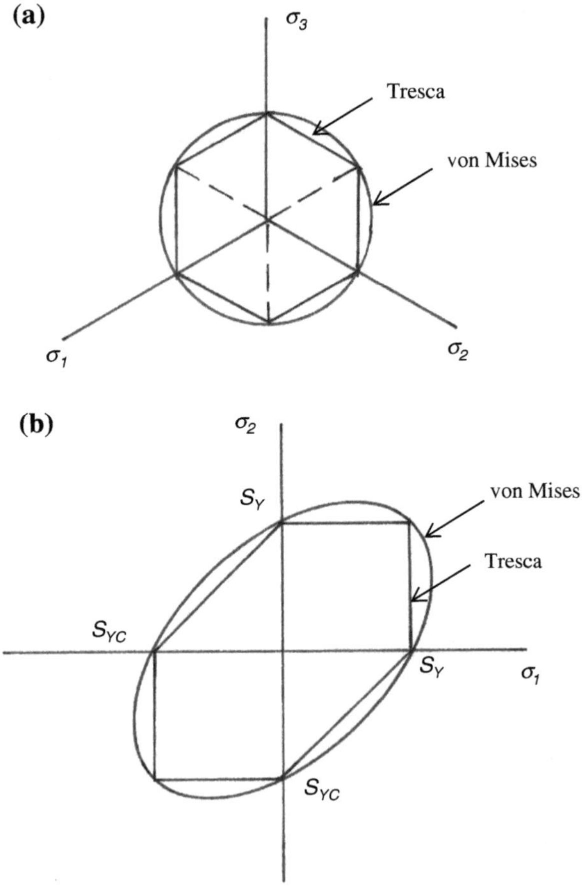

Fig. 4.8 Comparison of yield loci for von Mises and Tresca criterion. **a** Yield surface projection onto the π-plane, and **b** yield locus in the principal plane stress space

4.3 Plasticity and Hardening

Continuous loading beyond initial yield causes the material to deform with combined elastic and plastic strain components. Plasticity is characterized by a time-independent irreversible straining that can be sustained by a material upon reaching a certain level of stress. However, under high temperature loading, metals exhibit both creep and viscoplastic or time-dependent plastic deformation. The creep response of metals is discussed in Sect. 4.4. The separation of strains into plastic, viscoplastic and creep components is done primarily for analytical convenience since the strain components cannot be distinguished physically. Mathematical formulations of classic plasticity and viscoplasticity theory for implementation in FE analysis are well described in many textbooks (e.g. [22, 42]). A unified inelastic strain constitutive model that introduces an inelastic strain term to represent combined effects of plasticity, viscoplasticity and creep is described in Sect. 4.5.

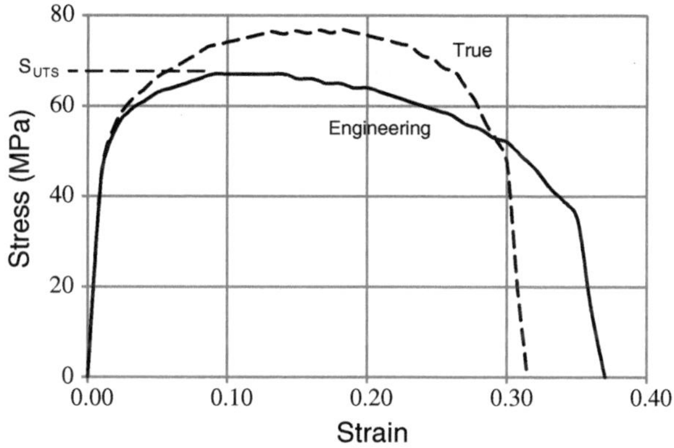

Fig. 4.9 Comparison of engineering and true stress–strain curves of SAC405 solder alloy at a testing strain rate of 19.74 s^{-1}. Data is extracted from Su et al. [51]

This section discusses physical aspects of the stress–strain diagram in relation to accurate prediction of the material hardening response under the temperature and strain rate range of interest for solder joint reliability assessment.

A new relationship between stress and strain is required to describe the post-yield behavior of the material. A sample stress–strain diagram for the whole range of measured strains during a tension test of SAC405 solder alloy at a specified strain rate is shown in Fig. 4.9. The measured data represents engineering stress–strain diagram (solid line). The stress is calculated as the measured force per unit original cross-sectional area, while the strain is obtained directly from an extensometer initially placed over the gage length of the test specimen. The true stress of the material within the gage volume should account for the continuous decrease in cross section throughout the tensile loading of the specimen. The engineering stress, S, and engineering strain, e, data can be converted into true stress, σ, and true or logarithmic strain, ε, respectively, using the following equations:

$$\sigma = S(1 + e) \tag{4.15}$$

$$\varepsilon = \ln(1 + e) \tag{4.16}$$

Equations (4.15) and (4.16) assume the conservation of volume during the deformation process. A true stress–strain curve (dashed line) is required for calculating the internal state of stresses and strains at each point in the material during FE analysis. At sufficiently large plastic strains, extensive necking occurs and the true stress at the necked section continuously increases. Such localized necking is associated with a triaxial stress state that renders the calculated true stress–strain curve at large inelastic strain inaccurate. However, the magnitude of inelastic strain accumulated in the critical solder joint of a BGA assembly at fracture is relatively small because failure often occurs due to a fatigue damage mechanism. Therefore, an additional criterion is required for predicting the fracture of the bulk solder joint. In this respect, the damage mechanics-based approach in predicting the fracture process is discussed in Chap. 7.

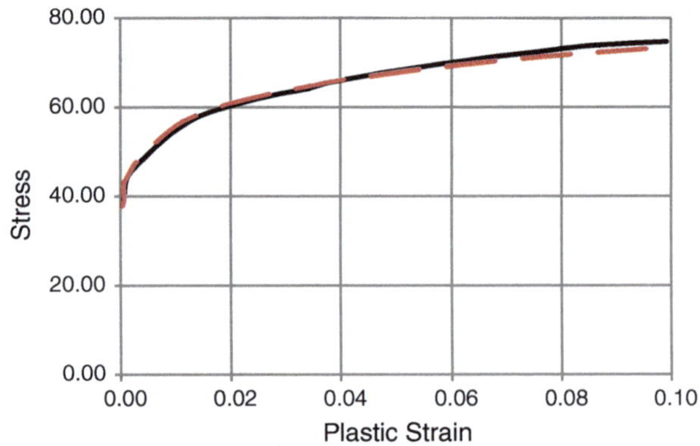

Fig. 4.10 Representation of true stress–plastic strain curve of SAC405 solder alloy for plastic strain up to 0.10. Dashed line represents Eq. (4.17)

Strain hardening of a solder alloy is manifested in a non-linear increase in strength with increasing strain of the material. This hardening characteristic for a SAC405 solder is illustrated in Fig. 4.10. The initial yield strength is 38 MPa. The stress–strain relationship in the plastic deformation range can be described using an analytical equation (also known as Ramberg-Osgood equation):

$$\varepsilon_p = K\sigma^m \tag{4.17}$$

The coefficient and exponent of the least-squared fit curve is $K = 1.4567 \times 10^{-17}$ and $m = 8.49$, respectively. The coefficient K can be temperature-dependent while exponent m is a material parameter. In many cases, Eq. (4.17) is inadequate in representing both temperature and strain rate sensitivity of the stress-plastic strain curves of the alloy. In this respect, the unified inelastic strain model can be employed to represent the inelastic response of the material to different combinations of temperature and strain rate. One such unified constitutive model (Anand model) is used in numerous FE simulation examples discussed in this book. Alternately, the stress–plastic strain curves of the material can also be introduced into FE analysis software by the piecewise linear representation of the curves, given the experimental data pairs. It is worth mentioning that constitutive models only calculate the stress–strain response to loading, but do not predict terminal fracture of the material. Consequently, a separate criterion for fracture is required, as discussed in Chap. 7, within the damage mechanics framework.

Hardening of a material can be visualized by progressive development of the yield surface with the subsequent yield surface being dependent on the degree of plastic straining. The degree of hardening, κ, can be related to a measure of total plastic deformation or effective plastic strain, $\bar{\varepsilon}_p$, integrated over the strain path as:

$$\kappa = \bar{\varepsilon}_p = \int d\bar{\varepsilon}_p \tag{4.18}$$

The incremental effective plastic strain can be defined in terms of deviatoric strain components:

$$d\bar{\varepsilon}_p = \sqrt{\left(\frac{2}{3}\{d\varepsilon'\}_p\{d\varepsilon'\}_p\right)} \tag{4.19}$$

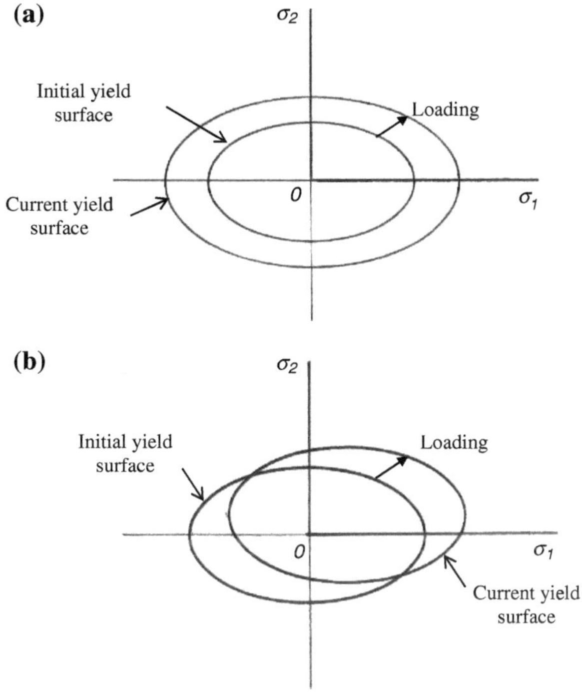

Fig. 4.11 Models for representation of strain hardening behavior. **a** Isotropic strain hardening and **b** kinematic strain hardening

The corresponding effective stress, $\bar{\sigma}$, is defined in terms of deviatoric stress, $\{\sigma'\}$ as:

$$\bar{\sigma} = \sqrt{\frac{3}{2}\{\sigma'\}\{\sigma'\}} \tag{4.20}$$

The strain hardening behavior is represented by the uniform expansion of the initial yield surface and referred to as an isotropic strain-hardening model, as illustrated in Fig. 4.11a. In a kinematic strain-hardening model, the new yield surface translates in the stress space while preserving its shape and orientation. The kinematic strain-hardening model is able to capture the Bauschinger effect, as displayed in many ductile materials during unloading from a plastic strain state.

The definition of stress level, stress and strain increment for a typical strain hardening material is illustrated using an uniaxial stress–strain diagram, as illustrated in Fig. 4.12. In this case, the effective stress, $\bar{\sigma}$, and effective plastic strain increment, $d\bar{\varepsilon}_p$, given by Eqs. (4.20) and (4.19), is reduced to σ and $d\varepsilon_p$, respectively. Elastic behavior up to the initial yield strength level, S_Y, is characterized by the elastic modulus, E. Subsequent elastic–plastic behavior is characterized by the hardening law, Eq. (4.18) in the following form:

$$\bar{\sigma} = H\left(\bar{\varepsilon}_p\right) \tag{4.21}$$

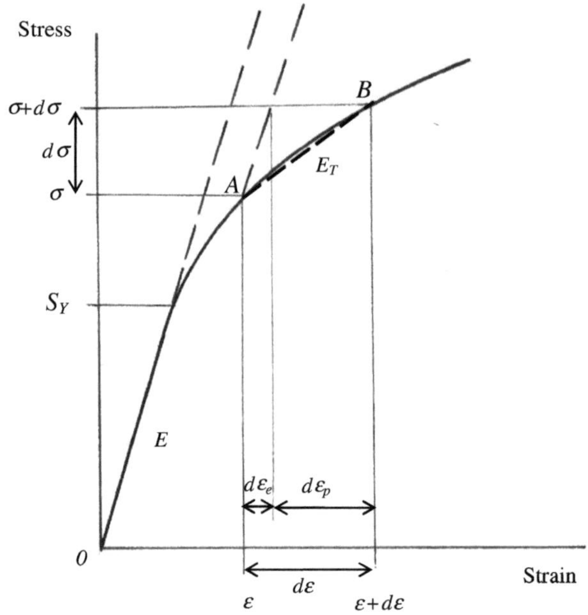

Fig. 4.12 Definition of elastic–plastic strain hardening variables for uniaxial case

Assuming that the current stress is denoted by point *A*, an increment of total strain, $d\varepsilon$, would correspond to the converged stress increment, $d\sigma$, and bring the stress to a new level at point *B*. The local gradient of the curve is defined as the elastic–plastic tangent modulus, E_T. The hardening function, obtained by differentiating Eq. (4.21), is written as:

$$H'(\bar{\varepsilon}_p) = \frac{d\sigma}{d\varepsilon_p} = \frac{E_T}{1 - \frac{E_T}{E}} \tag{4.22}$$

It is noted that the hardening function can be determined from the prescribed stress–plastic strain curve of the material. An example is given by Eq. (4.17) and illustrated in Fig. 4.10. A small total strain increment, $d\varepsilon$ (from small applied load increment), is desirable to ensure that the hardening function accurately represents the continuously varying slope of the hardening curve.

4.4 Creep Response

Solder joints in microelectronic components are often subjected to stress at high temperature, both during reliability testing and operating conditions. At high homologous temperatures, defined as T/T_m, where T_m is the melting temperature, solder alloys experience both phenomena of creep and viscoplasticity. Creep refers to the redistribution of stresses and strains with time in the material under

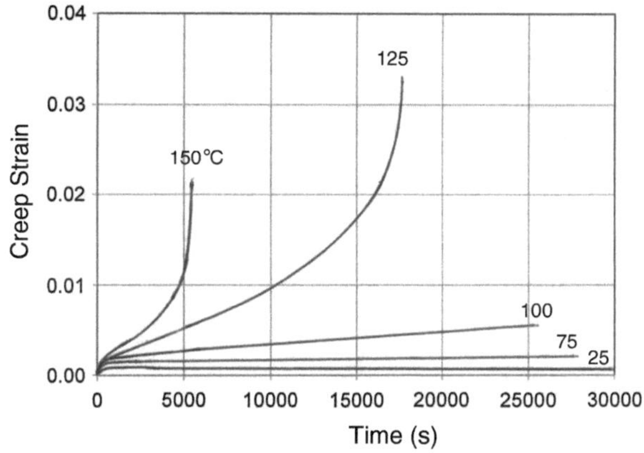

Fig. 4.13 Creep curves of as-reflowed SAC405 solder alloys at an applied stress of 10 MPa and different isothermal test temperature levels. Data extracted from Ma [32]

constant stress levels, while viscoplasticity is time-dependent plastic deformation. Creep and viscoplastic deformation are physically indistinguishable. The separate description of each phenomenon is for analytical convenience. Creep deformation of metals and alloys becomes significant above a homologous temperature of 0.4. Since reliability temperature cycles for packages with lead-free solder joints typically cover temperature ranges up to 150 °C, creep deformation is of great concern in assessing the reliability of these solder joints. At a peak temperature cycle of 150 °C, the homologous temperature of SAC405 solder is 0.68.

Creep characteristics of the solder alloy are determined through a creep test [5]. In this test, a constant load is applied to the creep specimen at isothermal test temperature levels and the resulting strain evolution is recorded until the specimen ruptures. Sample creep-rupture curves for SAC405 solders at an applied stress level of 10 MPa and different test temperature levels are shown in Fig. 4.13. An instantaneous elastic strain of ε_o is recorded upon loading. The strain initially evolves at a decreasing rate in the *Primary creep stage* into a normally steady-state *Secondary creep stage.* However, the assumption of constant creep rate during the secondary stage is only approximate for the SAC405 solder, as indicated in the figure, particularly at elevated temperature tests. At large creep strain levels, necking begins and the specimen deforms at an accelerated strain rate in the *Tertiary creep stage* to the final rupture of the material. Higher isothermal test temperatures (or applied stress) results in a faster creep rate and shorter rupture life of the material.

Several creep models have been introduced and examined to describe the steady-state creep response of solders [3, 17, 21, 25, 36]. One creep model that considers the diffusion creep mechanism, commonly employed for predicting the steady-state creep response of solder alloys, is expressed in power-law form as [36]:

$$\dot{\varepsilon}_{cr} = A\sigma^n \exp\left(-\frac{Q}{RT}\right) \tag{4.23}$$

where A and n are material constants, Q is the activation energy for creep process, and R and T are the universal gas constant and absolute temperature (in Kelvin

Table 4.2 Published creep model parameter values for several lead-free solder alloys

Solder alloy	Parameters			Source of data
	A (s^{-1})	n	Q (kJ/mol)	
(a) Power law creep model, Eq. (4.23)				
Sn–3.5Ag	1.5×10^{-5}	11.3	79.5	Mavoori et al. [35]
	5×10^{-6}	11	79.8	Wiese et al. [56]
	2×10^{-6}	11	73.2	Wiese et al. [59]
	6×10^{-25}	20	74.9	Wiese et al. [59]
Sn–4.0Ag	1×10^{-2}	9.6	94	Neu et al. [38]
Sn–3.8Ag–0.7Cu	6×10^{-23}	19	84.2	Wiese et al. [59]
	1×10^{-12}	13	75.2	Wiese et al. [59]
Sn–4.0Ag–0.5Cu	2×10^{-21}	18	83.1	Wiese and Rzepka [58]
	8×10^{-11}	12	74.8	Wiese and Rzepka [58]
	5.09×10^{-3}	6.27	76.2	Ma [32]
Sn–2.5Ag–0.8Cu–0.5Sb	1.45×10^{-10}	14	95	Neu et al. [38]
Sn–9.0Zn	2.17×10^{-2}	5.7	65.2	Mavoori et al. [35]

Solder alloy	Parameters				Source of data
	B (s^{-1})	β (MPa^{-1})	m	Q (kJ/mol)	
(b) Garofalo hyperbolic sine law, Eq. (4.24)					
Sn–3.5Ag	178.6	0.115	4.75	57.1	Wiese et al. [56]
Sn–3.8Ag 1.0Cu	0.0026	0.185	3	38.7	Wiese et al. [56]
Sn–3.8Ag–0.7Cu	3.2×10^4	0.037	5.1	65.3	Pang et al. [43]
Sn–3.9Ag–0.6Cu	4.41×10^5	0.005	4.2	45	Lau et al. [29, 30]
	248.4	0.188	3.79	62.3	Zhang et al. [66]
	0.184	0.221	2.89	62.0	Xiao and Armstrong [62]
	1.5×10^3	0.19	4	71.3	Zhang [67]
Sn–4.0Ag–0.5Cu	1.77×10^5	0.0548	4.89	76.13	Ma [32]
Sn–1.3Ag–0.2Cu–0.05Ni	8×10^5	0.1	7	75	Wiese et al. [59]
Sn–1.3Ag–0.5Cu–0.05Ni	6×10^4	0.1111	7	70	Wiese et al. [59]
Sn–2.7Ag–0.4Cu–0.05Ni	2×10^7	0.1429	5	100	Wiese et al. [59]

scale), respectively. Parameter values for this model are listed in Table 4.2a for several lead-free solder alloys. It is noted that several sets of model parameter values are made available for the same material, as obtained from the least-squared fit of the steady-state creep rate data. Such variation could have come from variations in grain size and structure of the creep test samples.

Another model for the steady-state creep stage is expressed in terms of the shear strain rate, $\dot{\gamma}$ and shear stress, τ in hyperbolic sine law as [17]:

$$\dot{\gamma} = B[\sinh(\beta\tau)]^m \exp\left(-\frac{Q}{RT}\right) \tag{4.24}$$

where B and m are material constants, and β is the stress multiplier. The hyperbolic sine creep model is able to represent the creep response of a material at low stress where the steady-state strain rate appears to be proportional to the power of applied stress until high strain rates are reached, and such proportionality breaks

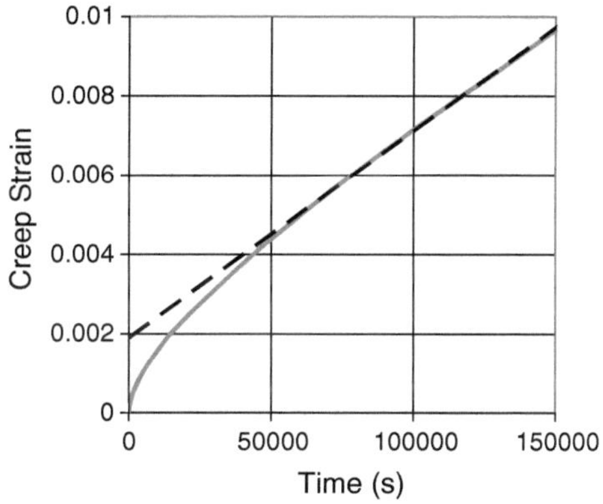

Fig. 4.14 Predicted creep curve for SAC405 solder at 14.7 MPa, 70 °C illustrates the primary and secondary creep stages. *Dashed line* represents measured data for steady-state creep rate from Wiese et al. [57]

down. Parameter values for this model are listed in Table 4.2b for several lead-free solder alloys.

During the reliability temperature cycling of electronic packages with BGA solder joints, both temperature and thermally induced stress in the solder joints varies during the temperature ramp that contributes primarily to viscoplastic deformation and fatigue of the solder. Creep deformation is dominant during the dwell time at peak high and low temperature levels. In view of the cyclic nature with a dwell time of the reliability temperature cycles, both primary and steady state stages of creep are significant when assessing the reliability of the solder joints. Thus, a prediction of time-to-rupture or rupture life of the solder alloys under isothermal and/or iso-stress conditions is not applicable in assessing the reliability of solder joints in BGA assembly. In this respect, the unified inelastic strain model is better suited for predicting the combined effects of viscoplasticity and creep. While details of one such model (Anand model) are discussed in Sect. 4.4, the model prediction of uniaxial creep of SAC405 solder at 14.7 MPa, 70 °C is illustrated in Fig. 4.14. Both primary and steady state creep stages are predicted. The dashed line represents the measured steady-state creep rate of the alloy creep tested under identical conditions [57].

4.5 Unified Inelastic Strain Theory

The separation of the inelastic strain component into time-independent plasticity, time-dependent (visco) plasticity and creep, as done in the classical mechanics approach, often leads to over-predicting the strain magnitude and inconsistencies in the final results [31]. Alternately, a unified inelastic strain model that recons similar underlying mechanisms of plasticity and creep represents both phenomena by a single constitutive equation with evolving state variables. Several unified inelastic strain models have been introduced (e.g. [3, 7, 10]). The Anand

Table 4.3 Description of parameters for the Anand unified inelastic strain model

Parameter	Description	Units
s_0	Initial value of state variable, s	MPa
Q/R	Activation energy term	K
A	Pre-exponential factor	s^{-1}
ξ	Stress multiplier	–
m	Strain rate sensitivity of stress	–
h_0	Hardening coefficient	MPa
$\hat{s}$	Coefficient for deformation resistance saturation value	MPa
n	Strain rate sensitivity of the saturation value	–
a	Strain rate sensitivity of the hardening coefficient	–

model, introduced for modeling the deformation of metals under high homologous temperatures, has been widely used to represent the deformation of solder joints during reflow and subsequent reliability temperature cycles. For brevity, only the Anand model is discussed in this book. The model is briefly introduced in Sect. 3.6.

The Anand model does not employ any explicit yield conditions for the loading-unloading criterion. Plastic strain is determined by the current stress, internal state variables and temperature. In addition, plastic flow is assumed to occur at all non-zero stress levels. The flow equation is expressed in the time-rate form of inelastic strain as:

$$\dot{\varepsilon} = A \exp\left(-\frac{Q}{RT}\right)\left[\sinh\left(\xi\frac{\sigma}{s}\right)\right]^{1/m} \tag{4.25}$$

The parameters appearing in Eq. (4.25) are described in Table 4.3. It is noted that temperature-dependency is incorporated through the Arrhenius creep form while stress-dependency is represented by the hyperbolic sine creep form of the equation [17]. The evolution of the internal state variable, s, representing the resistance to plastic flow, is:

$$\dot{s} = \left\{h_0\left|1-\frac{s}{s^*}\right|^a \cdot sign\left(1-\frac{s}{s^*}\right)\right\} \cdot \dot{\varepsilon}_{in} \tag{4.26a}$$

with the saturation value s^* defined as:

$$s^* = \widehat{s}\left[\frac{\dot{\varepsilon}_P}{A}\exp\left(\frac{Q}{RT}\right)\right]^n \tag{4.26b}$$

The Anand model parameter values for several Sn–Pb and lead-free solder alloys are compiled in Table 4.4. Quite different parameter values are reported for small differences in chemical compositions of Sn–Ag–Cu alloys, indicating a strong dependency and sensitivity to the parameters of the composition and microstructure of the alloy. The performance of the model for Sn–40Pb and SAC405 solder alloys is illustrated through different FE simulation cases throughout this book.

Table 4.4 Anand model parameter values for several leaded and lead-free solder alloys

	A (s^{-1})	Q/R (K)	ξ	$\hat{s}$ (MPa)	m	n	s_0 (MPa)	h_0 (MPa)	a	References
Sn–3.5Ag	2.23×10^4	89,006	6	73.81	0.182	0.018	39.09	3,321.15	1.82	Cheng et al. [12], Yan et al. [63]
	1.63×10^6	1.41×10^4	1.61	11.99	0.13	0.017	7.72	5.87×10^4	2.09	Pei and Ou [44]
Sn–1.0Ag–0.5Cu	2.42×10^7	8,400	0.043	1.005	0.168	8.1×10^{-4}	30	3,162	1.59	Cheng et al. [12], Kim et al. [26]
Sn–2.0Ag–0.5Cu	500	8,500	4.3	28.7	0.16	0.04	6.6	6,100	1.3	Reinikainen et al. [47]
Sn–3.38Ag–0.84Cu	500	9,000	7.1	39.4	0.3	0.03	1.3	5,900	1.4	Reinikainen et al. [47]
Sn–3.5Ag–0.75Cu	4.61×10^6	8,400	0.038	1.04	0.162	4.6×10^{-3}	–	3,090	1.56	Cheng et al. [12]
Sn–3.8Ag–0.5Cu	49,601	13,982	13	34.71	0.36	0.02	16.31	800,000	2.18	Wang et al. [55]
Sn–3.8Ag–0.7Cu	107.65	7,619	59.36	86.28	4.03	0.0046	22.64	9,002	1.3	Rodgers et al. [48]
	4.61×10^6	8,400	4.6	14.37	0.162	0.038	40.16	3,090	1.56	Pei and Ou [44]
	9,450.6	10,041	1.1452	13.3372	0.1158	0.0402	21.57	133.803	0.1082	Pei and Ou [44]
Sn–4.0Ag–0.5Cu	325	10,561	10	42.1	0.32	0.02	20	8×10^5	2.57	Wang et al. [55]
Sn–40Pb	96,200	8,110	6×10^{-4}	6.90×10^{-3}	0.303	1×10^{-9}	6.90×10^{-3}	6.90×10^{-12}	1	Wu et al. [61]
	3.86×10^4	6,966	5	122.7	0.2845	0.002	5,040	30,110	1.8	Rajib et al. [45]
	142,051	9,727	73.82	120.4	1.523	6.84×10^{-3}	68.71	206,403	2.619	Koh [27]
	4,991	5,711	2.547	31.22	0.2305	0.1362	5.643	60,236	1.578	Koh [27]
Sn–37Pb	1.09×10^7	9,400	0.07	0.998	0.316	1.37×10^{-5}	–	2,884	1.26	Amagai et al. [2]
Sn–36Pb–2Ag	4×10^6	9,400	1.5	13.8	0.303	0.07	12.42	1,380	1.3	Amagai [1]
	8.49×10^6	9,400	0.065	0.99	0.322	6.71×10^{-4}	–	2,951	1.33	Amagai et al. [2]
Pb–5Sn–2.5Ag	1.05×10^5	11,000	7	41.63	0.241	0.002	33.07	1,432	1.3	Wilde et al. [60]

Table 4.5 Parameter values for the unified inelastic strain model (Anand model) with refined parameters for SAC405 solder alloy

Parameter	Value
A (1/ s-1)	7,128
Q/R (K)	10,561
m	0.8181
n	1.091×10^{-3}
$\hat{s}$ (MPa)	30.45
ξ	10
a	1.6
s_0 (MPa)	Eq. (4.27)
h_0 (MPa)	Eq. (4.28)

Parameter	a_0	a_1	a_2	a_3	a_4
s_0	25.996	−0.014772	7.386×10^{-6}	–	–
h_0	9.85×10^4	−880	2.2	-2.1×10^6	1.05×10^8

Selected Anand model parameters were modified and refined to account for the observed sensitivity of the solder deformation response to temperature and strain rates [11, 44, 55]. An example of such model parameter refinement for SAC405 is to assume that the initial value of the internal state variable, s_0 to be temperature-dependent, while the hardening coefficient, h_o to be functions of both temperature and strain rate in polynomial forms as follows [28]:

$$s_0 = 25.996 - 1.4772\left(10^{-2}\right)T + 7.386\left(10^{-6}\right)T^2 \tag{4.27}$$

$$h_0 = 9.85\left(10^4\right) - 880T + 2.2T^2 - 2.1\left(10^{-6}\right)\dot{\varepsilon}_p + 1.05\left(10^{-8}\right)\dot{\varepsilon}_p^2 \tag{4.28}$$

The coefficients in Eqs. (4.27) and (4.28) are determined for SAC405 solder using published experimental data of the solder specimens tested at temperatures varying from 25 to 150 °C and strain rates ranging from 10^{-5} to 10^{-3} s^{-1} [55]. The resulting set of Anand model parameter values with further refinement of the parameters, s_o and h_o for SAC405 solder alloy, are listed in Table 4.5. This set of model parameter values is employed in the FE simulations described in this book.

The performance of the Anand model in predicting uniaxial hardening behavior of a SAC405 solder under a range of test temperature levels and strain rates is illustrated in Figs. 4.15 and 4.16, respectively. The Anand model with the refined parameter values listed in Table 4.5 is employed. Predicted curves are compared with published test data [55]. The reasonably close comparison of predicted and measured stress–plastic strain curves provides validation of the Anand constitutive model for the SAC405 solder. Consequently, the material model was employed in all simulation cases illustrated in this book.

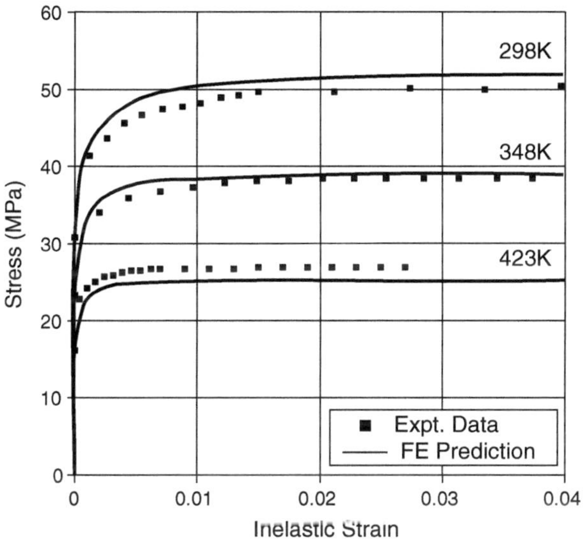

Fig. 4.15 Comparison of predicted and measured stress–inelastic strain curves for SAC405 solder at a strain rate of 10^{-3} s^{-1} and different test temperature levels. Experimental data are extracted from Wang et al. [55]

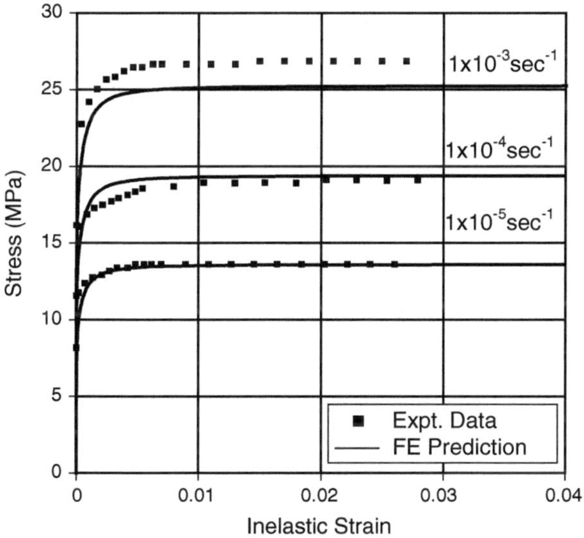

Fig. 4.16 Comparison of predicted and measured stress–inelastic strain curves for SAC405 solder at 150 °C (423 K) and different strain rates. Experimental data are extracted from Wang et al. [55]

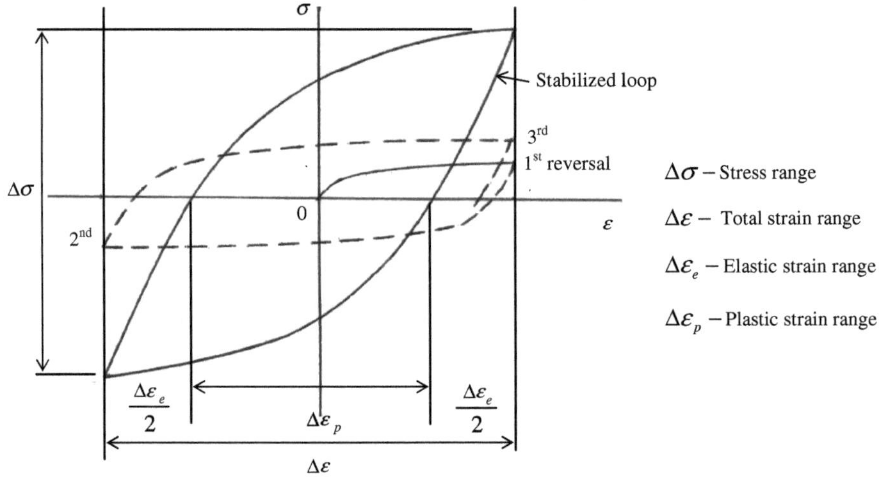

Fig. 4.17 Sketch of true stress–strain hysteresis loops for the first three load reversals for a cyclic hardening material and fatigue parameters defined for a stabilized loop

4.6 Cyclic Behavior

Temperature cycles for reliability testing of electronic packages consist of both temperature ramps and dwell time-periods. These temperature changes induce time-varying stresses in solder joints of tested BGA packages. The resulting stress cycles bring the solder into cyclic inelastic strain condition, such that the critical material point experiences low cycle fatigue. In this respect, a low cycle fatigue life model is required for predicting fatigue lives of the solder that, in turn, establish the reliability of the microelectronic assembly.

The low cycle fatigue behavior of metals and alloys is commonly described using uniaxial true stress–strain hysteresis loops obtained from a strain-controlled test [4]. A typical cyclic stress–strain curve for the first three reversals is schematically shown in Fig. 4.17 for a cyclic hardening material. Cyclic strain hardening is manifested in a higher magnitude of stress levels reached in subsequent load reversal. The stabilized hysteresis loop, often achieved after enduring a relatively large number of cycles, is also illustrated in the figure, along with a description of the relevant fatigue variables. Based on experimental data of hot-rolled steels, fatigue life (in terms of reversals to failure, *2N*) and total strain range, $\Delta\varepsilon$ can be expressed using the Manson–Coffin relationship [54]:

$$\frac{\Delta\varepsilon}{2} = \frac{\sigma'_F}{E}(2N)^b + \varepsilon'_F(2N)^c \tag{4.29}$$

Fatigue model parameters σ'_F, E, ε'_F, b and c are material constants extracted from low cycle fatigue test data.

The Coffin–Manson equation, Eq. (4.29), is employed in an updated form for fatigue life prediction of solder joints under reliability temperature and mechanical

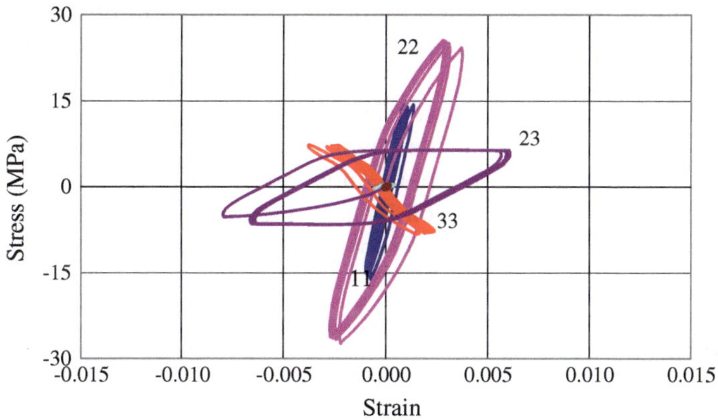

Fig. 4.18 Typical stress–strain hysteresis loops for selected Cartesian stress and strain components in the critical solder joint during mechanical cycles (11- and 33-direction is parallel to solder/IMC interface plane and 22-direction is normal to the interface)

load cycles. A complex stress state exists at each material point in the solder and undergoes cyclic variations. Typical true stress–strain hysteresis loops for selected Cartesian stress and strain components are shown in Fig. 4.18. These plots represent a low cycle fatigue response of a material point in the critical solder of a BGA package subjected to mechanical load cycles. It is noted that the shear component, τ_{23}, in the plane parallel to the solder/IMC interface displays the greatest total strain range. Thus, the inelastic shear strain range, $\Delta\gamma_{in}$, is expected to govern the fatigue life of the bulk solder joint. Based on the Coffin–Manson equation, the fatigue life model for the bulk solder joint is expressed as [34]:

$$N_f = c(\Delta\gamma_{in})^m \tag{4.30}$$

where the coefficient, c and exponent, m are determined experimentally. Other fatigue-related variables for use with Eq. (4.30) are the accumulated inelastic strain, $\varepsilon_{in,acc}$ and inelastic strain energy or work density, $W_{in,acc}$ per load cycle. Fatigue life model for viscoplastic solder is expressed in the form [13, 14]:

$$N_f = N_o + \frac{a}{da/dN} \tag{4.31}$$

where the parameter, a represents the characteristic crack length. The number of cycles to crack initiation, N_o and crack propagation rate, da/dN are both expressed as power-law function of the plastic work density, i.e.:

$$N_o = C_1\left(W_{in,acc}\right)^{m1} \tag{4.32}$$

$$\frac{da}{dN} = C_2\left(W_{in,acc}\right)^{m2} \tag{4.33}$$

The constants C_1, C_2, $m1$ and $m2$ are experimentally determined. The volume average of the inelastic work density accumulated per load cycle over elements across the critical solder/IMC interface plane is employed in FE prediction when using Eqs. (4.32) and (4.33) to reduce the sensitivity of the calculated variable to the size of the elements.

The predictive capability of these classical fatigue models for establishing reliability of microelectronic assemblies with BGA solder joints is illustrated with case studies, as elaborated in Chap. 6. Another alternative approach in reliability prediction of solder joints based on damage mechanics concept is discussed in Chap. 7, with illustrative examples in subsequent chapters.

References

1. Amagai, M.: Characterization of chip scale packaging materials. Microelectron. Reliab. **39**(9), 1365–1377 (1999)
2. Amagai, M., Watanabe, M., Omiya, M., Kishimoto, K., Shibuya, T.: Mechanical characterization of Sn–Ag-based lead-free solders. Microelectron. Reliab. **42**(6), 951–966 (2002)
3. Anand, L.: Constitutive equations for hot working of metals. Int. J. Plast. **1**(3), 213–231 (1985)
4. ASTM International: Standard Test Method for Strain-Controlled Fatigue Testing. ASTM E606 (2012)
5. ASTM International: Standard Test Methods for Conducting Creep, Creep-Rupture, and Stress-Rupture Tests of Metallic Materials. ASTM Standard E139 (2011)
6. ASTM International: Standard Test Methods for Tension Testing of Metallic Materials. ASTM E8/E8M-13a (2013)
7. Bodner, S.R., Partom, Y.: Constitutive equations for elastic–viscoplastic strain-hardening materials. ASME Trans. J. Appl. Mech. **42**, 385–389 (1975)
8. Bridgman, P.W.: Studies in Large Plastic Flow and Fracture. McGraw-Hill, New York (1952)
9. BS EN ISO 6892 2009-1: Metallic materials. Tensile testing. Method of test at ambient temperature
10. Chaboche, J.L.: Unified cyclic viscoplastic constitutive equations: development, capabilities, and thermodynamic framework. In: Krausz, A.S., Krausz, K. (eds.) Unified Constitutive Laws of Plastic Deformation, pp. 1–68. Academic Press, Waltham (1996)
11. Chen, X., Chen, G., Sakane, M.: Prediction of stress–strain relationship with an improved Anand constitutive model for lead-free solder Sn3.5-Ag. IEEE Trans. Compon. Packag. Technol. pp. 111–116 (2005)
12. Cheng, Z.N., Wang, G.Z., Chen, L., Wilde, J., Becker, K.: Viscoplastic Anand model for solder alloys and its application. Soldering Surf. Mt. Technol. **12**(2), 31–36 (2000)
13. Darveaux, R., Mawer, A.: Thermal and power cycling limits of plastic ball grid array (PBGA) assemblies. In: Proceedings of Surface Mount Technology Conference, pp. 315–326 (1995)
14. Darveaux, R.: Solder joint fatigue life model. In: Proceedings of Surface Mount Technology Conference, pp. 213–218 (1997)
15. Dutta, I., Pana, D., Marks, R.A., Jadhav, S.G.: Effect of thermo-mechanically induced microstructural coarsening on the evolution of creep response of SnAg-based microelectronic solders. Mater. Sci. Eng., A **410–411**, 48–52 (2005)
16. Dutta, I.: A constitutive model for creep of lead-free solders undergoing strain-enhanced microstructural coarsening: a first report. J. Electron. Mater. **32**(4), 201–207 (2003)
17. Garofalo, F.: Fundamentals of Creep and Creep-Rupture in Metals. The Macmillan Company, London (1966)
18. Gebhardt, E., Petzow, G.: The constitution of the Ag–Cu–Sn system Z. Metallkd. **50**, 597–605 (1959) (in German).

19. Glazer, J.: Microstructure and mechanical properties of Pb-free solder alloys for low-cost electronic assembly: a review. ASME Trans. J. Electron. Mater. **23**(8), 693–700 (1994)
20. Gong, J., Liu, C., Conway, P.P., Silberschmidt, V.V.: Modelling of Ag3Sn coarsening and its effect on creep of Sn–Ag eutectics. Mater. Sci. Eng. A **427**(1–2), 60–68 (2006)
21. Hacke, P., Sprecher, A.F., Conrad, H.: Computer simulation of thermo-mechanical fatigue of solder joints including microstructure coarsening. J. Electron. Packag. **115**(2), 153–158 (1993)
22. Huebner, K.H., Thornton, E.A., Byrom, T.G.: The Finite Element Method for Engineers. Wiley, New Jersey (1982)
23. Hwang, J.S.: Implementing Lead-Free Electronics. McGraw-Hill, New York (2004)
24. Japan Electronics and Information Technology Industries Association (JEITA): Lead-Free Roadmap 2002—Roadmap 2002 for Commercialization of Lead-Free Solder (2002)
25. Kashyap, B.P., Murty, G.S.: Experimental constitutive relations for the high temperature deformation of a Pb–Sn eutectic alloy. Mater. Sci. Eng. **50**(2), 205–213 (1981)
26. Kim, Y.B., Noguchi, H., Amagai, M.: Vibration fatigue reliability of BGA-IC package with Pb-free solder and Pb–Sn solder. In: Proceedings of 53rd Electronics Components and Technology Conference, pp. 891–897 (2003)
27. Koh, Y.K.: A Unified Constitutive Models for Solder Materials. Mechanical Engineering thesis, Universiti Teknologi Malaysia, Malaysia (2004)
28. Lai, Z.B.: Classical and Damage Mechanics-Based Models for Lead-Free Solder Interconnect. Master thesis, Universiti Teknologi Malaysia, Malaysia (2009)
29. Lau, J., Dauksher, W., Vianco, P.: Acceleration models, constitutive equations, and reliability of lead-free solders and joints. In: Proceedings of the 2003 Electronics Components and Technology Conference, pp. 229–236 (2003)
30. Lau, J.H., Wong, C.P., Lee, N.C., Lee, R.S.W.: Electronics Manufacturing with Lead-Free, Halogen-Free and Conductive-Adhesive Materials. McGraw-Hill, New York (2003)
31. Lee, W.W., Nguyen, L.T., Selvaduray, G.S.: Solder joint fatigue models: Review and applicability to chip scale packages. Microelectron. Reliab. **40**, 231–244 (2000)
32. Ma, H.: Characterization of Lead-Free Solders for Electronic Packaging. Ph.D. thesis, Auburn University, Alabama, USA (2007)
33. Ma, H.: Effects of temperature and strain rate on the mechanical properties of lead-free solders. J. Mater. Sci. **45**(9), 2351–2358 (2010)
34. Manson, S.S.: Thermal Stress and Low Cycle Fatigue. McGraw-Hill, New York (1966)
35. Mavoori, H., Chin, J., Vaynman, S., Moran, B., Keer, L., Fine, M.: Creep, stress relaxation, and plastic deformation in Sn–Ag and Sn–Zn eutectic solders. J. Electron. Mater. **26**(7), 783–790 (1997)
36. Mukherjee, A.K., Bird, J.E., Dorn, J.E.: Experimental correlation for high-temperature creep. Trans. Am. Soc Met. **62**, 155–179 (1969)
37. National Electronics Manufacturing Initiative (NEMI): Roadmap of Lead-Free Assembly in North America (2002)
38. Neu, R.W., Scott, D.T., Woodmansee, M.W.: Thermomechanical behavior of 96Sn–4Ag and castin alloy. ASME Trans. J. Electron. Packag. **123**(3), 238–246 (2001)
39. Nguyen, T.T., Yu, D., Park, S.B.: Characterization of the mechanical properties of actual SAC105, SAC405, and SAC405 solder joints by digital image correlation. Electron. Compon. Technol. Conf. **40**(6), 1409–1415 (2011)
40. Nimmo, K.: Alloy selections. In: Suganuma, K. (ed.) Lead-Free Soldering in Electronics-Science, Technology, and Environmental Impact, pp. 49–90. Marcel Dekker, New York (2004)
41. Ourdjini, A., Hanim, M.A.A., Koh, S.F.J., Siti Aisha, I., Tan, K.S., Chin, Y.T.: Effect of solder volume on interfacial reactions between eutectic Sn–Pb and Sn–Ag–Cu solders and Ni(P)–Au surface finish. In: 31st International Conference on Electronics Manufacturing and Technology, pp. 437–442 (2007)
42. Owen, D.R.J., Hinton, E.: Finite Elements in Plasticity: Theory and Practice. Pineridge Press Ltd., Swansea (1980)

43. Pang, J.H.L., Xiong, B.S., Low, T.H.: Creep and fatigue characterization of lead free 95.5Sn–3.8Ag–0.7Cu solder. In: Proceedings of the 2004 Electronic Components and Technology Conference, pp. 1333–1337 (2004)
44. Pei, M., Ou, J.: Constitutive modeling of lead-free solders. In: Proceedings of International Symposium on Advanced Packaging Materials: Processes, Properties and Interface, pp. 44–49 (2005)
45. Rajib, R., Raghuram, V.P., Suresh, K.S.: Thermal Cycle Guidelines for Automotive, Computer, Portable and Implantable Medical Device Application. ASME, New York (2000)
46. Rao, S., Bath, J., Ladhar, H.: Lead-free manufacturing. In: Lead-Free Electronics, pp. 101–138. Wiley, New Jersey (2006)
47. Reinikainen, T.O., Marjamaki, P., Kivilahti, J.K.: Deformation characteristics and microstructural evolution of SnAgCu solder joints. Proc. EuroSimE **2005**, 91–98 (2005)
48. Rodgers, B., Flood, B., Punch, J., Waldron, F.: Experimental determination and finite element model validation of the Anand viscoplasticity model constants for SnAgCu. Proc. EuroSimE **2005**, 490–496 (2005)
49. Sarah, L.A., Notis, M.R., Chromik, R.R., Vinci, R.P.: Microstructural evolution in lead-free solder alloys: Part I. Cast Sn–Ag–Cu eutectic. J. Mater. Res. **19**(5), 1417–1424 (2004)
50. Sidhu, R.S., Deng, X., Chawla, N.: Microstructure characterization and creep behavior of Pb-free Sn-rich solder alloys: Part II. Creep behavior of bulk solder and solder/copper joints. Metall. Mater. Trans. A **39**(2), 349–362 (2007)
51. Su, Y.A., Tan, L.B., Tee, T.Y., Tan, V.B.C.: Rate-dependent properties of Sn–Ag–Cu based lead-free solder joints for WLCSP. Microelectron. Reliab. **50**, 564–576 (2010)
52. Suh, D., Kim, D.W., Liu, P., Kim, H., Weninger, J.A., Kumar, C.M., Prasad, A., Grimsley, B.W., Tejada, H.B.: Effects of Ag content on fracture resistance of Sn–Ag–Cu lead-free solders under high-strain rate conditions. Mater. Sci. Eng., A **460–461**, 595–603 (2007)
53. Tai, S.F., Ourdjini, A., Khong, Y.L., Venkatesh, V.C.: Tamin, M.N.: Effect of phosphorus content and solid state aging on intermetallic formation between lead-free Sn–Ag–Cu solder and electroless nickel/immersion gold under bump metallurgy. In: Proceedings of the 4th International Symposium on Electronic Materials and Packaging, pp. 267–270 (2002)
54. Tavernelli, J.F., Coffin Jr, L.F.: Experimental support for generalized equation predicting low cycle fatigue. Trans. ASME J. Basic Eng. **84**(4), 533 (1962)
55. Wang, Q., Liang, L.H., Chen, X.F., Weng, X.H., Yong, L., Irving, S., Luk, T.: Experimental determination and modification of Anand model constants for Pb-free material 95.5Sn4.0Ag0.5Cu. In: Proceedings of EuroSimE 2007, pp. 1–9 (2007)
56. Wiese, S., Schubert, A., Walter, H., Dudek, R., Feustel, F., Meusel, E., Michel, B.: Constitutive behavior of lead-free solders vs. lead-containing solders-experiments on bulk specimens and flip-chip joints. In: Proceedings of the 2001 Electronic Components and Technology Conference, pp. 890–902 (2001)
57. Wiese, S., Meusel, E., Wolter, K.-J.: Microstructural dependence of constitutive properties of eutectic SnAg and SnAgCu solders. In: Proceedings of the 2003 Electronic Components and Technology Conference, pp. 197–206 (2003)
58. Wiese, S., Rzepka, S.: Time-independent elastic–plastic behaviour of solder materials. Microelectron. Reliab. **44**(12), 1893–1900 (2004)
59. Wiese, S., Roellig, M., Mueller, M., Wolter, K.-J.: The effect of downscaling the dimensions of solder interconnects on their creep properties. Microelectron. Reliab. **48**(6), 843–850 (2008)
60. Wilde, J., Becker, K., Thoben, M., Blum, W., Jupitz, T., Wang, G., Cheng, Z.: Rate dependent constitutive relations based on Anand model for 92.5Pb5Sn2.5Ag solder. IEEE Trans. Adv. Packag. **23**(3), 408–414 (2000)
61. Wu, C.M.L., Lai, J.K.L., Wu, Y.: Thermal–mechanical interface crack behavior of a surface mount solder joint. Finite Elem. Anal. Des. **30**(1–2), 19–30 (1998)
62. Xiao, Q., Armstrong, W.D.: tensile creep and microstructural characterization of bulk Sn3.9Ag0.6Cu lead-free solder. J. Electron. Mater. **34**(2), 196–211 (2005)

63. Yan, D.J., Huang, D.J., Huang, C.Y., Wu, Z.H., Zhou, X.: The reliability research of lead-free solder joint of flip-chip. In: Proceedings of 7th International Conference on Electronic Packaging Technology, ICEPT'06, pp. 1–5 (2006)
64. Yeh, C.-L., Lai, Y.-S.: Transient fracturing of solder joints subjected to displacement-controlled impact loads. Microelectron. Reliab. **46**(5–6), 885–895 (2006)
65. Zahn, B.A.: solder joint fatigue life model methodology for 63Sn37Pb and 95.5Sn4Ag0.5Cu materials. In: Proceedings of the 2003 Electronic Components and Technology Conference, pp. 83–94 (2003)
66. Zhang, Q., Dasgupta, A., Haswell, P.: Viscoplastic constitutive properties and energy-partitioning model of lead-free Sn3.9Ag0.6Cu solder alloy. In: Proceedings of the 2003 Electronics Components and Technology Conference, 1862–1868 (2003)
67. Zhang, Q.: Isothermal mechanical and thermo-mechanical durability characterization of selected Pb-Free solders. Ph.D. Thesis, University of Maryland, College Park, Maryland (2004)

Chapter 5
Application I: Solder Joint Reflow Process

5.1 Introduction

The solder joint reflow process exposes the test assembly with BGA solders and/or printed solder paste to prescribed temperature profiles resulting in complete melting, wetting and flowing of the molted solder. During reflow cooling, the solder solidifies, making up the solder joint. Higher reflow temperature also activates a chemical reaction at the solder/pad interface, resulting in the formation of a thin intermetallic layer that improves the strength of the interface. The large temperature difference between solder reflow and ambient conditions induces thermal strains and stresses in the solder joint due to mismatches in the coefficient of thermal expansion (CTE) among the different materials making up the assembly. Thus, a thorough understanding of the evolution of internal states in the solder joint is essential in mitigating the residual stresses in the solder joints and ensuring high reliability of the assembly. In this respect, finite element (FE) simulation with an appropriate model for the geometry of the assembly and suitable material models can be employed to describe the mechanics of the materials during the solder joint reflow cooling process.

This chapter illustrates the synthesis of the problem in establishing internal states of displacement, strain and stress in the critical solder joint of a typical test assembly with BGA solder joints during the solder reflow cooling process. The mathematical model describes aspects of physical representation of the problem for consideration in FE simulation. Essential requirements for FE modeling, as discussed in Chap. 3 are reiterated, but in the scope of the current problem. The FE simulation results are presented and discussed with respect to a thermally-induced warpage of the assembly, distribution of stresses, and inelastic strains in the critical solder joint and their potential effects on reliability and characteristic evolution of equivalent stress and inelastic strain with respect to potential premature failure of the test assembly.

M. N. Tamin and N. M. Shaffiar, *Solder Joint Reliability Assessment*,
Advanced Structured Materials 37, DOI: 10.1007/978-3-319-00092-3_5,

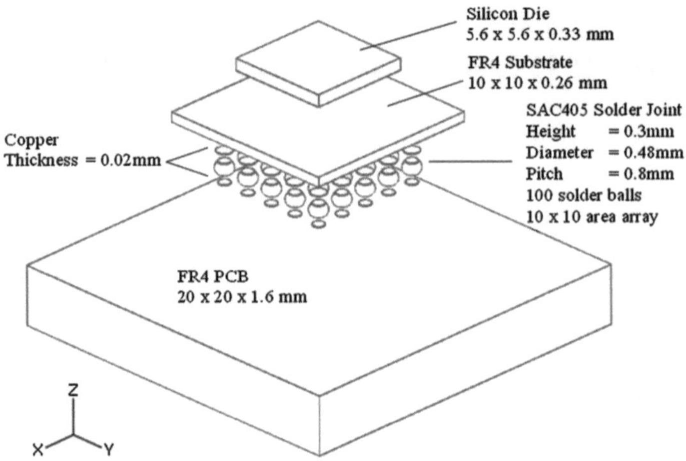

Fig. 5.1 Exploded view of the quarter model test assembly with BGA solder joints

5.2 Finite Element Modeling

The solder reflow cooling process of a typical microelectronic assembly with BGA solder joints is considered in this case study. The assembly consists of 100 SAC405 solder joints arranged in a square 10 × 10 solder array providing package interconnection to the PCB. The diameter, stand-off height and pitch distance of the solder joints is 480, 300 and 800 μm, respectively. All solder joints are modeled as Solder Mask-Defined (SMD) at the device or substrate side and Non-Solder Mask-Define (NSMD) at the PCB side. Due to the symmetry of the assembly geometry, boundary conditions and temperature loading, only one-quarter of the assembly is considered in FE modeling. An exploded view of the quarter model geometry of the test assembly is illustrated in Fig. 5.1, indicating various parts that make up the assembly. These include the silicon die representing the flip-chip package on the FR-4 substrate, copper and Cu_6Sn_5 IMC layers, and FR-4 PCB. The uniform thin layer (2 μm) of Cu_6Sn_5 phase is assumed to have formed upon reflow cooling.

Boundary conditions for the quarter model representing the *x*-*z* and *y*-*z* symmetry of the test assembly are identical to those described in Sect. 3.3. The reflow temperature for SAC405 solder joints is taken at 220 °C where the assembly is assumed to be in a stress-free condition and perfectly flat. The assembly is then cooled to 25 °C at the prescribed cooling rate of 80 °C/min. In addition, all surfaces initially in contact are assumed to be perfectly bonded throughout the reflow cooling process. It is further assumed that the transient heat transfer effect is negligible and that all parts of the model assembly experience the same rate of cooling.

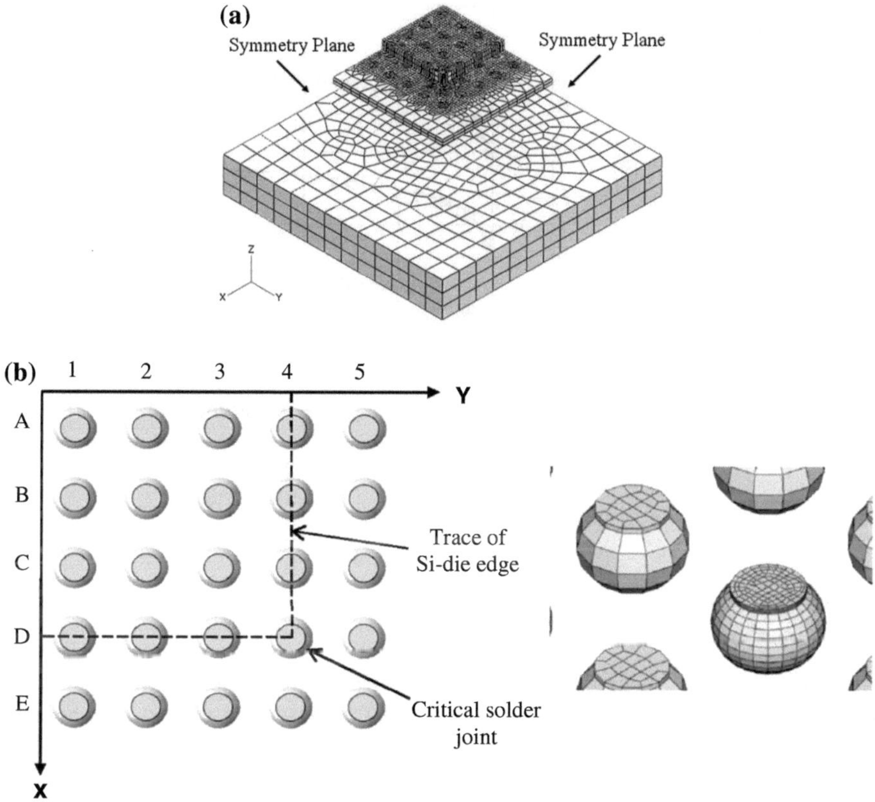

Fig. 5.2 **a** Comparison of the different mesh sizes employed for solder joints and other parts of the test assembly. **b** The fine element mesh of the critical solder joint

A mesh convergence study was performed focusing on BGA solder joints of the assembly. The procedures are discussed in Sect. 3.2. The resulting mesh sizes for all parts of the assembly used in the analysis are compared in Fig. 5.2a. Based on a preliminary simulation of the model assembly with a coarse element mesh size of the BGA solder joints, the most critical solder joint with respect to thermally induced strain was identified to be the one located underneath the corner of the Si-die (flip chip), as illustrated in Fig. 5.2b. This critical solder joint is then discretized into a finer element mesh for subsequent convergence analysis. The finest size of an element employed in this critical solder joint has an edge length of 0.01 mm. This fine-size element mesh ensures that the predicted FE results of the internal states in the critical solder joint are independent of the size of the elements used in the analysis. In order to reduce the number of unknown degrees of freedoms for this simulation, thus reducing computational time and resources, only the critical solder joint is discretized into fine element mesh.

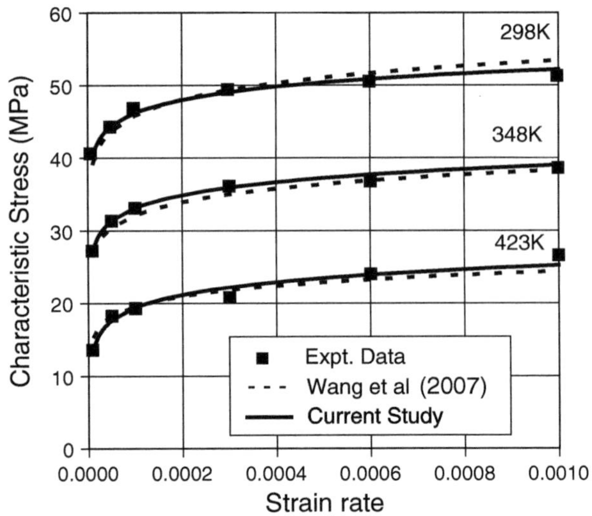

Fig. 5.3 Variation of characteristic stress levels with strain rates for a SAC405 solder alloy at different test temperatures. Test data and dashed lines are extracted from Wang et al. [9]

5.2.1 Constitutive Model for SAC405 Solder

The stress-strain response for a SAC405 solder alloy is adequately represented by the unified inelastic strain model (Anand model), as described in Sect. 4.5. Selected Anand model parameters, namely the initial value of the state variable, s_0, and the hardening coefficient, h_0, are further refined to account for the observed sensitivity of the alloy deformation response to temperature gradients and strain rates [see Eqs. (4.27) and (4.28)]. The performance of the model has also been examined, as shown in Figs. 4.15 and 4.16 for different temperature levels and strain rates. It is worth noting that the predictive capability of the model has been examined and validated for an inelastic strain magnitude of up to 4 pct. Nevertheless, a good prediction of the stress-strain response for a strain magnitude of up to 10 pct. can be expected based on the identical trend of the true stress-strain diagram for SAC405, as illustrated in Fig. 4.9. Failure of the solder joint occurs due to fatigue and creep-fatigue mechanisms with limited plastic strain at fracture.

Additional validation of the constitutive model for the material can also be established by comparing other parameters pertaining to the stress-strain diagram. One such example is shown in Fig. 5.3, which compares the model prediction and measured characteristic stress levels corresponding to plastic strain level of 0.04 for SAC405 solder at different test temperatures and strain rates. Test data are extracted from Wang et al. [9]. Results show that the characteristic stress level reached in the solder increases from 41 MPa in a quasi-static tension to 52 MPa at 10^{-3} s^{-1} for tests conducted at 25 °C. However, the characteristic stress level reached is lower at higher temperature.

The Anand model parameters for a SAC405 solder, as listed in Table 4.5, was extracted from the set of experimental data covering temperatures and strain rates

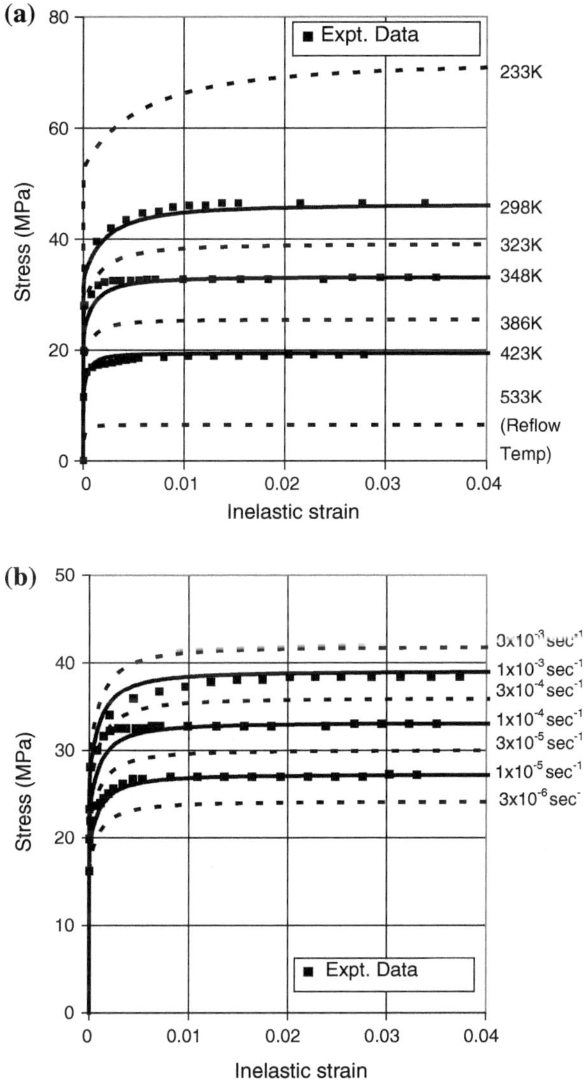

Fig. 5.4 Predicted tensile stress-inelastic strain curves of the SAC405 solder alloy **a** at a strain rate of $10^{-4}\ s^{-1}$ and different temperatures **b** at 75 °C and different strain rates. Test data extracted from Wang et al. [9]

from 25 to 150 °C and 10^{-5} to $10^{-3}\ s^{-1}$, respectively. However, the constitutive model will be employed in the simulation of solder reflow cooling, with temperatures ranging from 220 to 25 °C. In addition, an electronic package or assembly will also be subjected to a sub-zero temperature level of −40 °C during a typical reliability temperature cycle. Consequently, the performance of the model to predict the stress-strain curves outside the range that was used to determine their parameter values should be examined. The predicted stress-strain curves at various temperature levels and strain rates are shown in Fig. 5.4. Solid lines represent predicted responses at the respective temperature and strain rate corresponding to the

Table 5.1 Properties of various materials that make up the assembly, and are employed in FE simulation

Material	E (GPa)	ν	α (ppm/°C)
SAC405	44.7	0.36	20
Wang et al. [9], Schubert et al. [6]	−0.146 T(°C)		
Si-die	132.46	0.28	2.113
Shi et al. [7]	−0.00954 T(K)		+0.00235 T(K)
Cu layer	141.92	0.35	15.64
Shi et al. [7]	−0.0442 T(K)		+0.0041 T(K)
Cu_6Sn_5	85.6	0.31	16
Amagai [1]			

ν_{xy}	ν_{yz} and ν_{xz}	α_x and α_y ppm/°C)	α_z (ppm/°C)
FR4 PCB and substrate (isotropic plane x-y) [10]			
0.11	0.39	16	84
E_x and E_y (MPa)	E_z (MPa)	G_{xy} (MPa)	G_{xz} and G_{yz} (MPa)
27,924	12,204	12,600	5,500
−37 T(K)	−16 T(K)	−16.7 T(K)	−7.3 T(K)

experimental data, while the dashed lines are predicted responses at other temperature levels. The results demonstrate the capability of the material model to interpolate and extrapolate stress-strain curves with consistent trends for different test temperatures and strain rates.

Other materials that make up the assembly are assumed to behave elastically throughout the reflow cooling process. Temperature-dependent properties of these materials are expressed as linear variations of the test temperature, as summarized in Table 5.1. In addition, orthotropic behavior of the FR4 substrate and PCB is also considered.

5.3 Residual Internal States in the Critical Solder Joint

Finite element results for a simulation of solder reflow cooling of the assembly with BGA solder joints are presented and discussed in terms of the predicted warpage of the assembly, evolution characteristics of stress and strain in the critical solder joint throughout the reflow cooling process, and residual stress and inelastic strain distribution in the solder at the end of the solder reflow cooling process.

5.3.1 Reflow Process-induced Warpage of the Assembly

The cooling process from a reflow temperature of 220 °C to ambient level of 25 °C exposes the assembly with BGA solder joints to a temperature change of $\Delta T = -195$ °C. Mismatches in coefficient of thermal expansion (CTE) of the

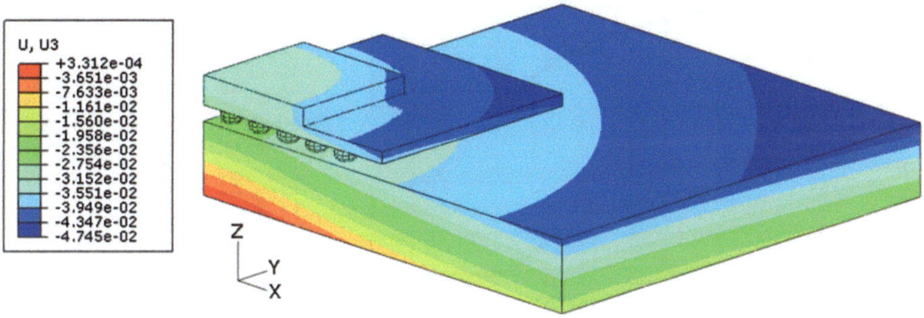

Fig. 5.5 Warpage of the quarter model assembly showing a vertical displacement field at 25 °C following the solder reflow process

various materials making up the assembly induce thermal strains and corresponding stresses in the materials. The constraint of perfectly bonded surfaces that are initially in contact resulted in warpage of the assembly following the solder reflow cooling process. Figure 5.5 shows the predicted deformation or warpage of the quarter model of the assembly with a convex shape (*sad face*) at 25 °C. The FR4 PCB and substrate with a CTE value of 16 ppm/°C would have greater free contraction during reflow cooling compared to the Si-die with a CTE value of 3 ppm/°C. However, since both layers remain bonded, the board and substrate pull the Si-die and deform it into a convex shape during reflow cooling. The magnitude of warpage, given by the maximum out-of-plane deformation of the assembly, is 212 μm.

A deformation field in the warped assembly is further analyzed by examining the out-of-plane deformation of the Si-die, substrate and PCB along a chosen diagonal path of the assembly, as illustrated in Fig. 5.6. Results show that both Si-die and PCB experience a smooth downward deflection. Deformation of the substrate region below the Si-die is dictated by the stiffer Si-die material, while a sudden change in the deformation profile occurred in the region away from the corner of the die where the stiff constraint by the Si-die diminishes. It is obvious from the deformed configuration that the solder joint (denoted as D4) located below the corner of the Si-die would be subjected to the greatest constraint, thus experiencing higher stress. This critical solder joint will continue to accumulate greater inelastic strain during the subsequent temperature cycling of the assembly. A similar location of the critical solder joint has also been identified for FC-PBGA assembly with 18 × 18 area array solder joints [8] and another assembly with 7 × 17 area array solder joints subjected to cyclic temperature loading [3].

The deformed shape of the row of solder joints along the diagonal plane of the quarter model of the assembly at 25 °C following the solder reflow cooling process is shown in Fig. 5.7. The magnitude of displacement is magnified by 20X to give a clearer visual of the relative deformation. The significant difference in curvature of the warped substrate and PCB resulted in the solder joint D4 to experience the greatest deformation. The magnitude of thermally induced strains and stresses in the solder joints exceeded the initial yield conditions of the SAC405 solder alloy.

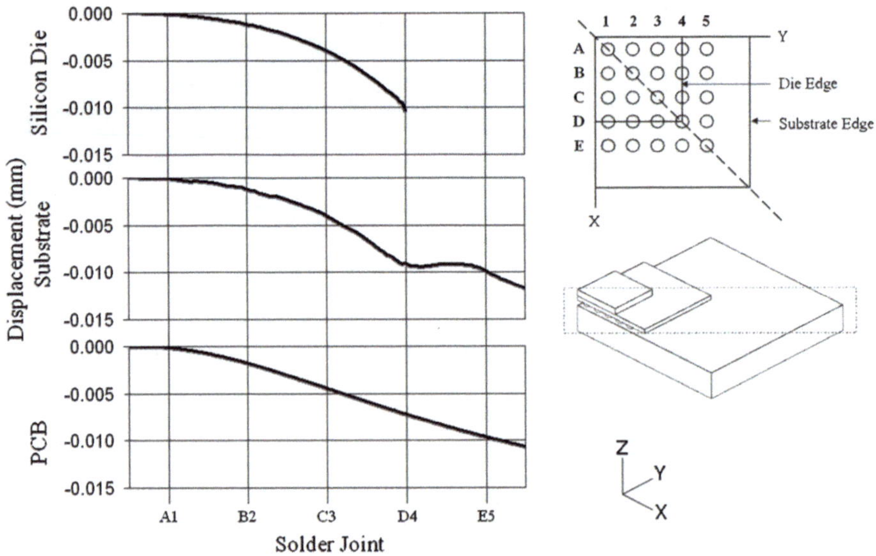

Fig. 5.6 Out-of-plane deformation of the Si-die, substrate and PCB along the diagonal path of the assembly, following the solder reflow cooling process

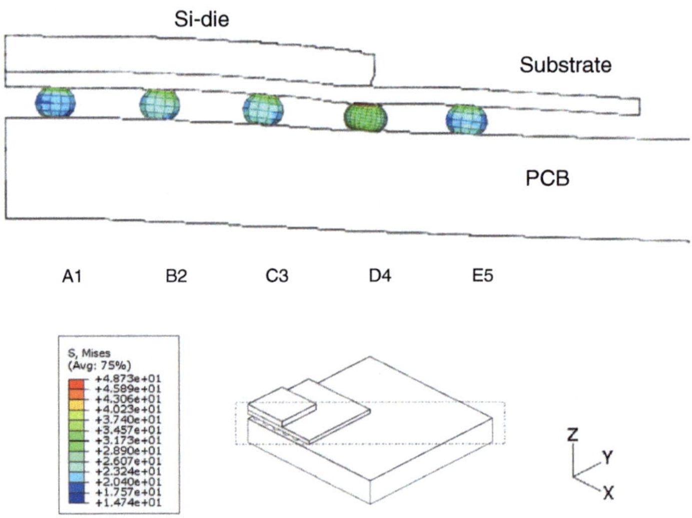

Fig. 5.7 The deformed row of solder joints along the diagonal plane of the assembly at 25 °C following the solder reflow cooling process

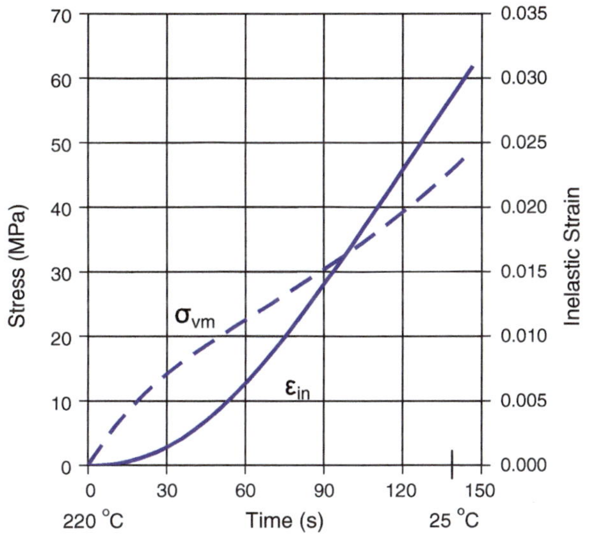

Fig. 5.8 Evolution of von Mises stress and equivalent inelastic strain on the critical SAC405 solder joint throughout solder reflow cooling process

5.3.2 Evolution of Residual Stresses and Strains

The predicted evolution of inelastic strains and the corresponding von Mises stress at a typical material point in the critical solder joint throughout the reflow cooling process is shown in Fig. 5.8. Results show that inelastic strain develops early following the solder reflow process and continuously evolves throughout the cooling process. Although the stress is low at this high temperature level, the yield strength of the SAC405 solder alloy is also relatively low, as previously depicted in Fig. 5.4. The local strain rate in the material also increases to a constant magnitude of $2.9 \times 10^{-4}\ s^{-1}$ below the temperature of 86.7 °C that corresponds to the elapsed cooling time of 65 s, as indicated in the figure. Inelastic strain accumulates throughout the cooling process with the corresponding stress brought to the yield surface. The residual von Mises stress reaches the saturation stress level of 48.7 MPa (refer also to Fig. 5.4 for the strain rate of $2.9 \times 10^{-4}\ s^{-1}$), with a corresponding inelastic strain of 3.10 pct. at 25 °C. The accumulated inelastic strains consist primarily of a creep strain component in view of the high homologous temperature, T/T_m, of the SAC405 solder alloy. In addition, the magnitude of residual strains and stresses in the solder is affected by the cooling rate during the cooling stage of the reflow temperature profile. A rapid cooling rate will induce a greater magnitude of inelastic strain due to a higher intrinsic hardening rate of the alloy at a higher straining rate. However, a too-long cooling time period will subject the solder joints to higher temperatures over a longer duration that, in turn, accumulates higher inelastic (creep)

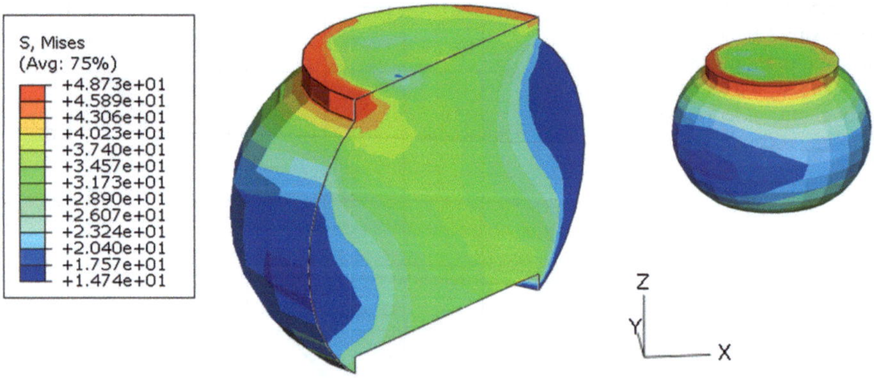

Fig. 5.9 von Mises stress distribution in the critical solder joint at 25 °C following solder reflow cooling. Stress scale is in MPa. The section plane lies at 45° to the *x*-axis

strains. Thus, an optimized cooling rate is desired. The inelastic strain will continue to accumulate during the subsequent temperature cycling of the assembly to potentially cause the initiation of crack by the low-cycle fatigue mechanism.

5.3.3 Residual Stress and Strain Fields

The distribution of von Mises stress in the critical solder joint at 25 °C following the solder reflow cooling process is shown in Fig. 5.9. The stress scale is in MPa. The stress field indicates a localized region of high stress and a stress gradient at the edge of the solder joint near the SAC405/Cu_6Sn_5 interface. This is due to the severe CTE mismatch between adjacent materials and the sharp edge of the SMD geometry. Such a relatively high stress level favors localized yielding of the solder joint. The yield stress of the SAC405 solder alloy at 25 °C is about 19 MPa at the strain rate of 10^{-5} s^{-1} [3]. A high stress region is predicted near the solder/IMC interface on the component (top) side of the assembly. This result is consistent with observed fatigue cracking that preferentially propagates in the solder near the component side in thermal cycle tests [4]. A relatively higher stress magnitude is developed in a narrow central region across the diagonal section of the bulk solder joint. Such stress distribution is a consequence of relative planar (*x*–*y* plane) displacement between the substrate and PCB of the assembly at SMD and NSMD section, respectively. In subsequent temperature cycles, this region will become a favorable path for fatigue crack propagation under a shear-dominated fracture. This aspect of damage propagation in the bulk solder joint is addressed in Chap. 9 of this book.

The corresponding equivalent inelastic strain field in the critical solder joint is shown in Fig. 5.10. An inelastic strain concentration is confined to a small edge region and shallow depth from the free surface near the solder/IMC interface. The remaining bulk solder material remains elastic. This highly localized inelastic region

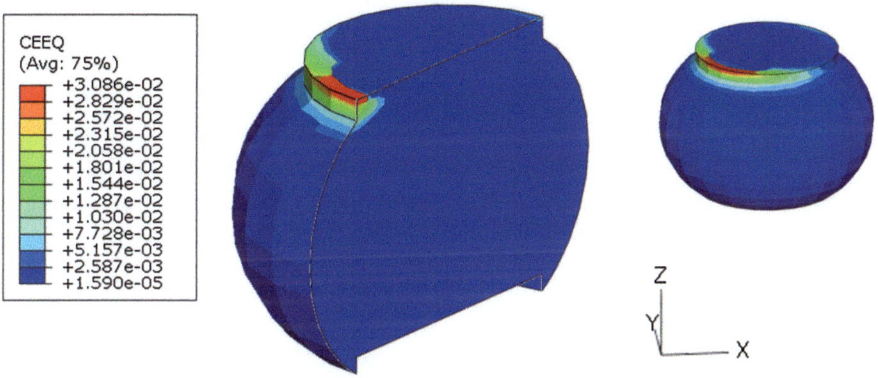

Fig. 5.10 Equivalent inelastic strain distribution in the critical solder joint at 25 °C following solder reflow cooling. The x–z section plane lies at 45° to the x-axis

at the periphery of the solder/IMC interface will likely serve as a favorable site for fatigue crack initiation during subsequent reliability temperature and/or mechanical cycling of the assembly. A similar crack initiation site corresponding to the location of the maximum creep and plastic strain in the solder has been established [5].

References

1. Amagai, M.: Chip scale package (CSP) solder joint reliability and modeling. Microelectron. Reliab. **39**(4), 463–477 (1999)
2. Ma, H.: Characterization of lead-free solders for electronic packaging. Ph.D. Thesis, Auburn University, Alabama (2007)
3. Park, S.B., Ahmed, I.Z.: Shorter field life in power cycling for organic packages. ASME Trans. J. Electron. Packag. **129**(1), 28–34 (2007)
4. Rooney, D.T., Castello, N.T., Cibulsky, M., Abbott, D., Xie, D.: Materials characterization of the effect of mechanical bending on area array package interconnects. Microelectron. Reliab. **44**(2), 275–285 (2004)
5. Sarihan, V.: Temperature dependent viscoplastic simulation of controlled collapse solder joint under thermal cycling. ASME Trans. J. Electron. Packag. **115**(1), 16–21 (1993)
6. Schubert, A., Dudek, R., Auerswald, E., Gollhardt, A., Michel, B., Reichl, H.: Fatigue life models for snagcu and snpb solder joints evaluated by experiments and simulation. In: Proceedings of the 2003 Electronic Components and Technology Conference, pp. 603–610 (2003)
7. Shi, X.Q., Wang, Z.P., Zhou, W., Pang, H.L.J., Yang, Q.J.: A new creep constitutive model for eutectic solder alloy. ASME Trans. J. Electron. Packag. **124**(2), 85–90 (2002)
8. Verma, K., Park, S.-B., Han, B., Ackerman, W.: On the design parameters of flip chip PBGA package assembly for optimum solder ball reliability. IEEE Trans. Compon. Packag. Technol. **24**(2), 300–307 (2001)
9. Wang, Q., Liang, L.H., Chen, X.F., Weng, X.H., Yong, L., Irving, S., Luk, T.: Experimental determination and modification of Anand model constants for Pb-free material 95.5Sn4.0Ag0.5Cu. In: Proceedings of EuroSimE 2007, pp. 1-9 (2007)
10. Zahn, B. A.: Solder joint fatigue life model methodology for 63Sn37Pb and 95.5Sn4Ag0.5Cu materials. In: Proceedings of the 2003 Electronic Components and Technology Conference, pp. 83–94 (2003)

Chapter 6
Application II: Solder Joints Under Temperature and Mechanical Load Cycles

6.1 Introduction

The reliability of microelectronic packages and assemblies is established through reliability testing of the device. In a reliability test, assemblies with BGA solder joints are subjected to temperature cycles, as prescribed by reliability test standards (e.g. [11]). The various different cyclic temperature profiles suggested for electronic packages and assemblies with solder interconnections are listed in Tables 3.1 and 3.2 (Chap. 3). Temperature loading parameters are peak temperature levels that set the temperature range, heating and cooling rates, and dwell-time period at peak temperature levels. The prescribed temperature profiles represent accelerated temperature loading for the reliability tests. Such loading produces similar characteristics of the actual stress factors experienced by the assembly during processing, reliability testing and service. These tests generate similar failure mechanisms in the solder joint as would otherwise be encountered under service conditions. The profile represents the expected temperature excursions experienced by the solder joints during service and power on-off of the device. The applied temperature range induces residual strains and stresses in the various materials of the assembly through mismatches in CTE, while the prolonged exposure to dwell-time periods at peak temperature levels promotes creep of the solder joint. Temperature cycling causes stress fluctuation in the solder joint while the heating and cooling rates of the temperature cycles dictate both a temperature- and a strain-rate dependent response in the material. The resulting cyclic stresses and strains induce low-cycle fatigue damage in the solder joint. Since solder joints are the weakest link in these assemblies, their reliability is dictated by solder joint reliability.

While temperature loading is inherent in the operation of a BGA assembly, mechanical loading is inevitable throughout the design life of numerous electronic devices containing such an assembly. These mechanical loadings include flexural load by pressing the keypads, accidental dropping of a mobile device, and mechanical vibrations induced during transportation and handling of the finished products. Therefore, the reliability of the solder joints should also be assessed under

M. N. Tamin and N. M. Shaffiar, *Solder Joint Reliability Assessment*,
Advanced Structured Materials 37, DOI: 10.1007/978-3-319-00092-3_6,

simulated and accelerated mechanical load cycles. In this respect, several mechanical tests on electronic assembly have been developed and introduced to investigate the integrity issues of the solder joints [9, 10]. Bending tests are useful for evaluating board-level assemblies for performance under mechanical stresses. Two types of bending tests performed on electronic packages and assemblies are the three-point-bend test [22, 24, 31] and the four-point-bend test [2, 7, 8, 12, 13, 17, 20, 26], with single or multiple electronic components and either single-sided or double-sided assemblies. A torsion test imparts cyclic mechanical twisting on a PCB with BGA package and is able to reproduce fatigue failure mechanisms in solder joints, as observed in a temperature cycling test [19]. An accelerated cyclic mechanical test is an effective and efficient method in generating solder joint reliability data that can be exploited in establishing fatigue life prediction models for the solder joints.

FE simulation has been employed in evaluating the fatigue performance and establishing the reliability of solder joints in electronics assemblies [21, 28, 32, 34]. The analysis provides better insight into the characteristic evolution of the internal states of the solder joints throughout the applied load cycles. A thorough understanding of the material response is essential in describing the failure process and assessing the reliability of solder joints. This chapter describes additional procedures considered in performing FE simulation and demonstrates the development of the fatigue life prediction model for the solder joints.

6.2 BGA Assembly Under Temperature Cycles

An FE simulation of the response of an electronic assembly with BGA solder joints under reliability temperature cycling is deliberated in this chapter. Additional aspects of FE modeling relevant to this problem, but not covered in previous chapters, are presented. Results are discussed in terms of the evolution of internal states and fatigue-related variables in the critical solder joint of the assembly. Fatigue life models based on this critical solder joint failure are developed.

6.2.1 Finite Element Modeling

A finite element model of a microelectronic assembly with BGA solder joints discussed in this case study is identical to that described in Chap. 5 (see Figs. 5.1, 5.2). The prescribed boundary conditions for the quarter model of the assembly are also applicable, as described in Sect. 3.4. In addition, an identical material model, properties and parameters for SAC405 solder joints are employed. In fact, the reflow cooling simulation can be continued with the application of subsequent temperature cycles. However, the effects of solder reflow process-induced stress and inelastic strain on subsequent thermal fatigue of solder joints has been investigated [15, 28]. The predicted shear strain hysteresis loops for stabilized fatigue

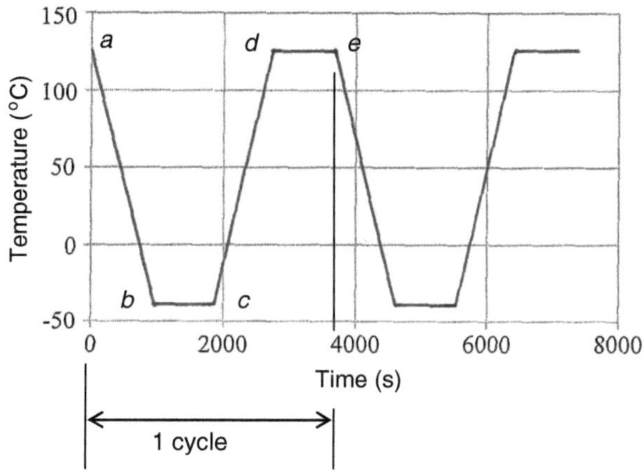

Fig. 6.1 Temperature profile used for simulation of solder joint reliability under temperature cycles

cycles for cases with and without simulated solder reflow cooling are identical. Thus, solder reflow process-induced stress and inelastic strain do not affect the fatigue life of solder joints in BGA assemblies. Consequently, for illustrative purposes, the effect of reliability temperature cycles on the evolution of internal states in the critical solder joint is examined with the initial stress-free condition corresponding to the peak temperature level of the cycle. The first two simulated reliability temperature cycles are illustrated in Fig. 6.1. The temperature cycles consist of a temperature range, $\Delta T = 165$ °C between 125 and −40 °C, with a heating and cooling ramp rate of 11 °C/min. A dwell time period, $t_{dwell} = 900$ s, is simulated at each peak temperature level. Thus, the temperature cycle is 60 min.

It is desirable to calculate several fatigue-related variables, such as the accumulated inelastic strain, $\varepsilon_{in,\,acc}$, inelastic shear strain range, $\Delta\gamma_{in}$ and accumulated inelastic work density per load cycle, $W_{in,acc}$, in the critical region of the critical solder joint. These variables are calculated as the volume-average quantities per cycle over elements at the critical location. The variables are used to establish fatigue life models of BGA assemblies based on the critical solder joint. A general form of several fatigue life models for the solder were discussed in Sect. 4.6. The critical location was identified based on solder reflow cooling analysis being at the solder/IMC interface region on the SMD side of the assembly, as discussed in Chap. 5. The region is discretized further into four layers of elements over a thickness of 25 μm, as shown in Fig. 6.2. Such refined mesh would help reduce sensitivity of the calculated volume-average quantities of fatigue-related variables to the size of the elements.

6.2.2 Evolution of Internal States in the Solder Joint

Evolution of von Mises stress and equivalent inelastic strain at a selected material point (marked *A* in Fig. 6.4) in the critical solder joint is plotted against cycle time, as shown in Fig. 6.3. In this thermal fatigue simulation, the assembly with BGA

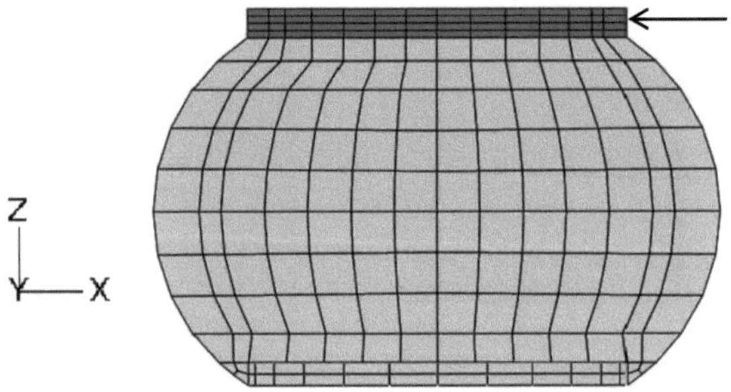

Fig. 6.2 Refined mesh in the solder mask-defined (*SMD*) interface region (*dark color*) for volume averaging calculations of fatigue-related variables

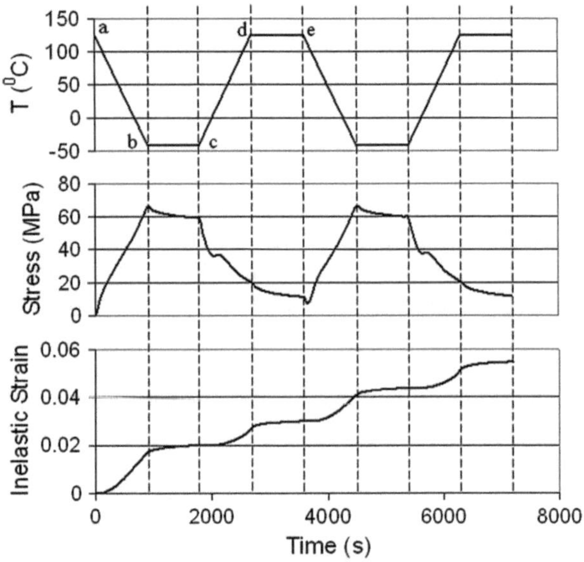

Fig. 6.3 Evolution of von Mises stress and equivalent inelastic strain in the critical solder joint during the first two simulated reliability temperature cycles

solder joints assumes a stress-free condition at the peak temperature of 125 °C (point *a*). Stress develops throughout the cooling temperature ramp to −40 °C (path *a–b*) due primarily to a mismatch in CTE of the solder and the adjacent IMC layer. The increase in stress is accompanied by the accumulation of inelastic strain in the solder over the cooling path of the cycle. Slight stress relaxation occurs over the 900-s dwell time period at −40 °C (path *b–c*) with the corresponding minute increase in the accumulated inelastic strain. This suggests that creep of the solder is operating with the highest stress magnitude in the material, even at such a low temperature. A further decrease in stress due to the effect of CTE mismatch occurs as the temperature ramps up to 125 °C (path *c–d*). A notable increase in the creep strain rate is predicted when the temperature reaches 40 °C (the corresponding

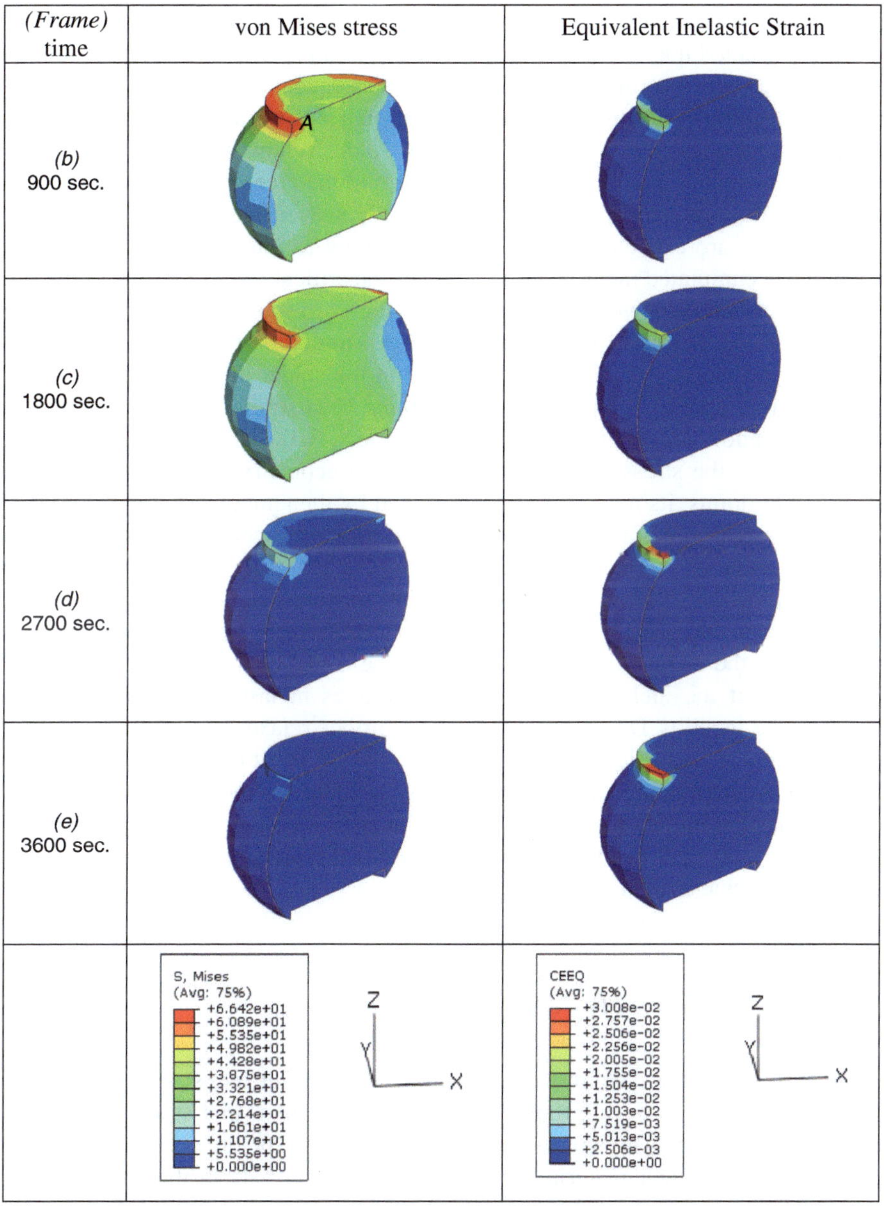

Fig. 6.4 Distributions of von Mises stresses and equivalent inelastic strains in the critical solder joint at selected time fames during the first temperature cycle. Stress scale is in MPa

stress is 35 MPa) along the heating ramp, likely due to an effective combination of viscoplasticity and creep deformation. A further increase in creep strain occurs over the dwell time period at the upper peak temperature (path *d*–*e*). This characteristic evolution of stress and strains in the solder joint repeats in subsequent temperature cycles. It is noted that the accumulated inelastic strain (0.031) is greater for the first temperature cycle than the second (0.024) and subsequent cycles.

The magnitude of stress and equivalent inelastic strain accumulated during temperature cycles is dictated by the initial solder reflow condition, the geometry of the assembly, hardening characteristics and the creep response of the solder alloy, and the prescribed temperature profile. However, identical mechanisms of solder failure are expected. Fatigue crack is likely to initiate at a material point when the accumulated inelastic strain reaches the critical level for crack initiation.

The distribution of von Mises stress and the corresponding equivalent inelastic strain in the critical solder joint at selected simulation time steps is shown in Fig. 6.4. The stress and inelastic strain field in the solder joint at −40 °C following the initial cooling temperature ramp (*frame b*) is similar to that predicted for the solder reflow process (see Figs. 5.9, 5.10, Sect. 5.3), but with a different magnitude. While the stress contour depends on the geometry of the assembly with BGA solder joints, its magnitude is determined by the temperature range and peak values. Limited stress relaxation occurred following dwell-time at a cold temperature (*frame c*). A subsequent heating temperature ramp resulted in significant stress relief in the solder joint, accompanied by extensive inelastic strain (*frame d*). Stress and inelastic strain are localized in a small edge region of the solder/IMC interface on the SMD side of the assembly. The localized stress relaxes with limited spread of the inelastic region and continuous inelastic strain accumulation in the same locality. Dwell time at 125 °C further relaxes the stress and subsequently induces inelastic (creep) strain at the critical point of the solder (*frame e*). Subsequent temperature cycles continuously accumulate inelastic strain and will likely cause the initiation of a fatigue crack at this critical location. Fatigue cracks in a solder joint have been observed to initiate and propagate from similar sites in chip scale packages [1, 14].

6.2.3 *Stress-Strain Hysteresis*

A general 3-D stress system occurs at any material point in a solder joint. As the temperature changes during the cycle, all Cartesian stress and strain components change accordingly. The stress-strain hysteresis loops for the six Cartesian components at the critical material point (point *A* in Fig. 6.4) in the critical solder joint is shown in Fig. 6.5. The shear stresses in direction-13, τ_{13} and direction-23, τ_{23} act in the plane of the solder/IMC interface, while normal stress in direction-33, σ_{33} acts in the direction normal to the interface plane. Since the critical region in the solder joint is located with a 45°-symmetry plane, as illustrated in the inset figure, the stress-strain components in direction-11 and direction-22 are identical, while shear stress-strain components in direction-13 and direction-23 are similar. The highest strain range is observed for the shear stress-strain components in direction-13 and direction-23. This is likely due to deformation

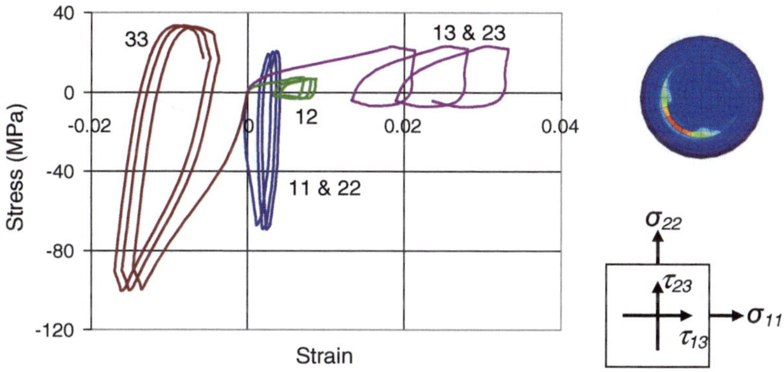

Fig. 6.5 Cartesian components of stress and strain at the critical location in the solder, plotted for the first three temperature cycles. The critical material point is located in the *red* colored region

(warpage) of the assembly and a CTE mismatch between the solder joint and the IMC when subjected to temperature cycles. The induced shear stress leads to excessive accumulation of plastic strain [14]. The large inelastic shear strain range is believed to be the primary contributor to solder joint fatigue. Thus, the parameter can be employed in fatigue life prediction of solder joints based on the Coffin-Manson strain-based model [16, 25]. The normal stress-strain component in direction-33 exhibits the largest stress range. The tensile part of this normal stress component is detrimental to an opening mode-I cracking along the solder/IMC interface.

The combined response of all Cartesian cyclic stress and strain components could also be described by plotting the hysteresis loop in terms of von Mises stress and equivalent strain. The von Mises stress is as described in Eq. (4.14b), while the corresponding equivalent strain is defined as:

$$\varepsilon_{eq} = \frac{\sqrt{2}}{3}\left[(\varepsilon_1 - \varepsilon_2)^2 + (\varepsilon_2 - \varepsilon_3)^2 + (\varepsilon_3 - \varepsilon_1)^2\right]^{\frac{1}{2}} \tag{6.1}$$

where ε_1, ε_2 and ε_3 is the principal strain in the respective principal direction. Figure 6.6 shows the hysteresis plot for the first simulated temperature cycle (path *a–b–c–d–e* in Fig. 6.1). The sign for von Mises stress is assigned to be positive for decreasing temperatures, where stress is induced through the CTE mismatch of the various materials in the assembly and negative for increasing temperatures. Results show that the critical material point in the solder joint experienced a von Mises stress magnitude of 66.4 MPa and the corresponding equivalent inelastic strain of 0.0183 when the temperature decreased from 125 to −40 °C (path *a–b*). Stress relaxation occurs during the dwell time period of 900 s with additional accumulated strain (path *b–c*). The stress is then plotted on the (−ve) scale for the heating stage. When the temperature is subsequently increased, the von Mises stress is reduced to 19.98 MPa with recoverable elastic strain to an equivalent

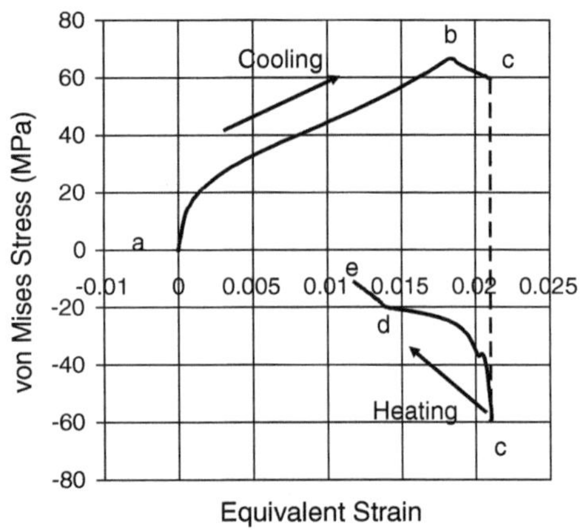

Fig. 6.6 Equivalent (von Mises) stress-strain hysteresis loop in the solder joint for the first temperature cycle with dwell time

strain of 0.0140, as indicated by path *c–d*. The stress relaxes further to 11.39 MPa at the end of the temperature cycle due to a high temperature creep effect during the dwell period (path *d–e*). These locked-in residual stresses and strains are available for the beginning of the next temperature cycle. It is noted that during heating between −40 and 25 °C, some stress components change sign due to the effect of bending induced by warpage of the assembly, as reflected in the inflection point along the hysteresis curve.

6.2.4 Fatigue Life Prediction Models

Phenomenological-based fatigue life models for microelectronic packages and assemblies have been discussed in Sect. 4.6. The fatigue life of the assembly is dictated by fatigue of the solder joint. The models correlate measured mean fatigue life of the assemblies, N_f with calculated characteristic fatigue-related variables of the solder joint in the assembly for a given reliability temperature cycle. In this chapter, two of the fatigue life models for assemblies with BGA solder joints subjected to temperature cycles are examined. The models employ the accumulated inelastic strain per cycle, $\varepsilon_{in,acc}$, and accumulated inelastic work density per cycle, $W_{in,acc}$, at the saturation hysteresis cycle as fatigue variables for correlation with the measured mean fatigue life of the assemblies. These chosen fatigue variables are extracted from FE simulation of the assembly with BGA solder joints subjected to different temperature cycles.

In the absence of measured mean life data for BGA assemblies and temperature cycles considered in this case study, the life data is generated using previously published fatigue models and treated as "measured" values for eight different combinations of geometry of the test assembly and temperature cycle. In this respect, two fatigue life models for SAC405 solder joints are considered [27]. One model employs a double power-law creep constitutive model for the

Table 6.1 Creep constitutive models with model parameter values for SAC405 solder

Creep	Constitutive model
Double power law creep [30]	$\dot{\varepsilon}_{cr} = A_1 \exp\left(-\frac{H_1}{kT}\right)\left(\frac{\sigma}{\sigma_n}\right)^{n_1} + A_2 \exp\left(-\frac{H_2}{kT}\right)\left(\frac{\sigma}{\sigma_n}\right)^{n_2}$
	$A_1 = 4 \times 10^{-7}\ \text{s}^{-1},\ H_1/k = 3{,}223,\ n_1 = 3.0,$
	$A_2 = 1 \times 10^{-12}\ \text{s}^{-1},\ H_2/k = 7{,}348,\ n_2 = 12,$
	$\sigma_n = 1$ MPa
	$E(\text{MPa}) = 59{,}533 - 66.667T(^\circ\text{K})$
Hyperbolic sine creep [23]	$\dot{\varepsilon}_{cr} = A_1[\sinh(a\sigma)]^n \exp\left(\frac{-H_1}{kT}\right)$
	$A_1 = 277{,}984\ \text{s}^{-1},\ a = 0.02447\ \text{MPa}^{-1},$
	$n = 6.41,\ H_1/k = 6{,}500$
	$E(\text{MPa}) = 61{,}251 - 58.5T(^\circ\text{K})$

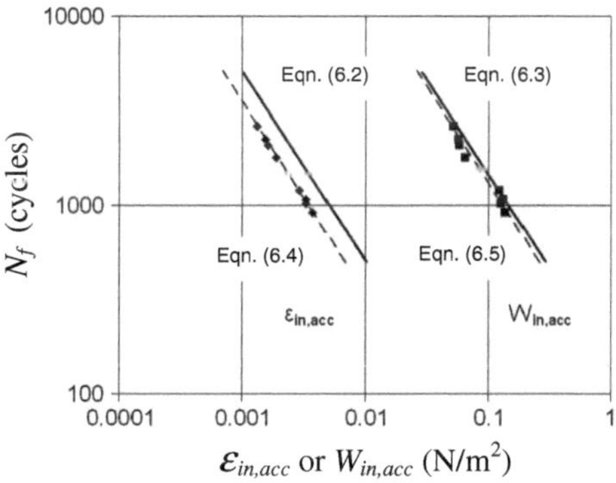

Fig. 6.7 Comparison of fatigue life models based on accumulated inelastic strain per cycle and accumulated inelastic work density per cycle

solder joint to calculate the characteristic accumulated inelastic strain per cycle, expressed as [30]:

$$N_f = (0.1968\ \varepsilon_{in,acc})^{-1} \tag{6.2}$$

The other utilizes a hyperbolic sine creep model for the solder to predict the characteristic accumulated inelastic work density per load cycle, expressed as [23]:

$$N_f = (0.0069\ W_{in,acc})^{-1} \tag{6.3}$$

The double power law creep and hyperbolic sine creep constitutive models along with parameter values for a SAC405 solder alloy used in simulating the "measured" life data of the solder joints are shown in Table 6.1. Both fatigue models (Eqs. 6.2 and 6.3) are represented by solid lines in Fig. 6.7.

The eight different combinations of electronic assemblies with BGA solder joints and reliability temperature cycles considered in developing fatigue models

Table 6.2 Details of electronic assembly models and temperature cycles employed in developing fatigue life models for SAC405 solder joints

Component	Assembly A		Assembly B
Si-die (mm)	5.6 × 5.6 × 0.33		6 × 6 × 0.7
Substrate (mm)	10 × 10 × 0.26		10 × 10 × 0.4
Solder array	10 × 10 area		8 × 8 area
Pitch (mm)	0.8		0.8
Solder stand-off height (mm)	0.3		0.3
Solder joint diameter (mm)	0.48		0.48
Load case id	ΔT (°C)	Rate (°C/min)	t_{dwell} (s)
Temperature cycles			
TC1	−40 to 125	11	900
TC2	−40 to 125	22	450
TC3	−50 to 125	14.5833	180
TC4	−50 to 125	29.1667	90

based on the life of SAC405 solder joints are summarized in Table 6.2. FE modeling procedures, as described in Chaps. 3, 5 and Sect. 6.2, are implemented where appropriate. The unified inelastic strain model (Anand model) is employed to predict the solder behavior in the simulations. The calculated fatigue variables, namely accumulated inelastic strain, $\varepsilon_{in,acc}$ and accumulated inelastic work density per load cycle, $W_{in,acc}$, for each case are correlated with a "measured" fatigue life of the assembly, N_f, as shown by solid symbols in Fig. 6.7. The resulting fatigue life models, curve-fitted with an identical exponent value of −1, can be expressed respectively as:

$$N_f = (0.2914\ \varepsilon_{in,acc})^{-1} \tag{6.4}$$

$$N_f = (0.0078\ W_{in,acc})^{-1} \tag{6.5}$$

The exponent of −1 is maintained to acknowledge the creep-dominated failure mechanism of the solder. Accumulated inelastic strain per cycle is shown to produce a better correlation to mean fatigue life of the solder as reflected in an excellent fit of the data points than the inelastic work density counterpart. Recall that solid lines represent the fatigue life prediction of the SAC405 solder joints in BGA assemblies when using the double power law creep and hyperbolic sine creep equations. Results indicate that the use of the unified inelastic strain model (Anand model) to represent the SAC405 solder response to temperature cycles leads to lower predicted lives compared to the double power law and hyperbolic sine creep models. Although different creep models affect the FE calculated variables, such as $\varepsilon_{in,acc}$ and $W_{in,acc}$, employed in the fatigue model, thus predicted fatigue lives, the relative trend of the fatigue performance is similar [18, 33]. The large number of measured life data on BGA assemblies of similar designs and a range of temperature cycles is necessary to validate the fatigue life model, thus offering greater confidence in the life prediction of solder joints.

Another characteristic fatigue variable, the inelastic shear strain range, $\Delta\gamma_{in}$, has been employed in predicting the fatigue lives of solder joints in view of its prominent magnitude among other Cartesian components in the hysteresis loops. The power-law form of this phenomenological-based model is shown in Eq. (4.30). The model has been evaluated for Sn-40Pb solder joints, as expressed in Sect. 6.4 [25].

6.3 BGA Assembly Under Cyclic Flexural Load

In this chapter, the response of a microelectronic assembly with BGA solder joints subjected to cyclic flexural loading is described. Stress-strain hysteresis in the critical solder joint of the assembly under the mechanical load cycles is discussed. Characteristic fatigue-related variables are then used in establishing a fatigue life model of the assembly.

6.3.1 Finite Element Modeling

The test assembly used in this simulation case study consists of a plastic ball grid array (PBGA) package surface mounted onto a PCB. Due to the symmetry of the geometry, boundary conditions and the four-point bend test set-up, only a quarter model of the set-up is considered for FE simulation. The geometry of the quarter model in the four-point bend test set-up is illustrated in Fig. 6.8a. The package is represented by a $10 \times 10 \times 0.3$ mm^3 Si-die with a 1.17 mm-thick over-mold. The 256 SAC405 solder joints connecting the package to the PCB are arranged in 4 peripheral rows, as illustrated in Fig. 6.8b. Solder joint diameter, standoff height, solder ball pitch distance and Cu-pad diameter are 932, 556, 1270 and 635 μm, respectively (Refer to Fig. 3.2, Chap. 3 for definition of dimensional parameters). Nickel and intermetallic $(Cu,Ni)_6Sn_5$ layers are modeled between the copper pad and solder joint. The four-point bend test set-up has an inner and outer span between *rigid* rollers of 100 and 135 mm, respectively. The loading would expose the package to tensile loading through bending of the PCB. Contact conditions between the PCB and rollers are assumed to be frictionless.

The symmetry boundary conditions of the quarter model of the test assembly are prescribed as follows (Refer to Fig. 6.8a):

Symmetry about the *x–z* plane:

$$U_y = U_{Rx} = U_{Rz} = 0 \tag{6.6}$$

Symmetry about the *y–z* plane:

$$U_x = U_{Ry} = U_{Rz} = 0 \tag{6.7}$$

The loading roller is allowed to move only in the *z*-direction to apply the load. Rotation about the *y*-axis is allowed. The corresponding boundary condition for the loading roller can be expressed as:

$$U_x = U_y = U_{Rx} = U_{Rz} = 0 \tag{6.8}$$

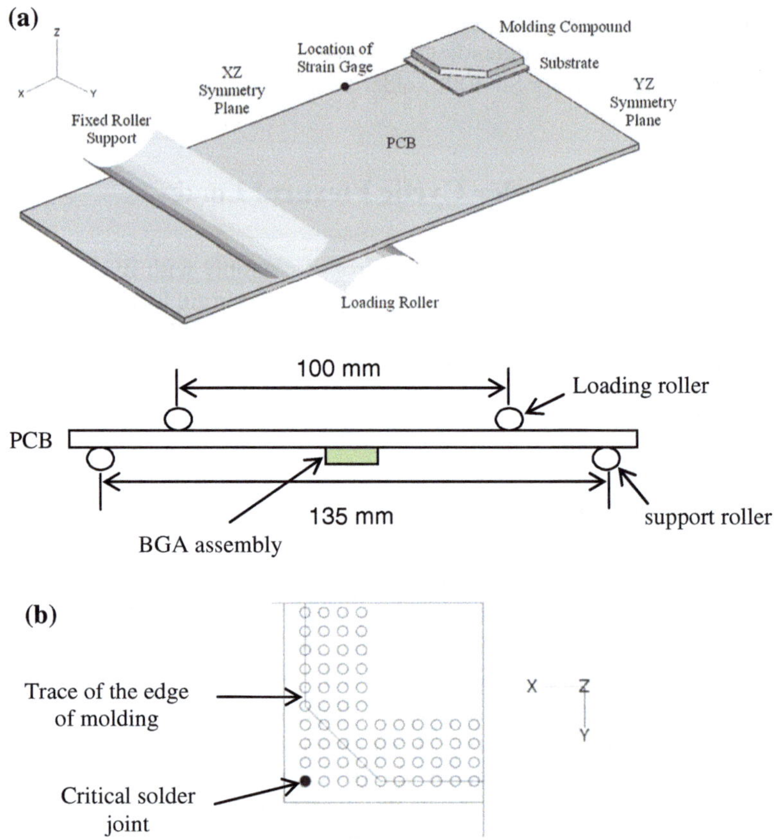

Fig. 6.8 **a** Quarter model of the test assembly with PBGA package in the four-point bend test set-up and **b** peripheral array of solder joints

The support roller is constrained from linear movement. However, rotation about the *y*-axis is allowed. These requirements are expressed by the following boundary conditions:

$$U_x = U_y = U_z = U_{Rx} = U_{Rz} = 0 \tag{6.9}$$

The FE model is simplified from that used in an experimental study on the reliability of the PBGA assembly under cyclic bending [12, 13]. This enables measured results to be compared with FE calculations for providing partial validation of the model.

The unified inelastic strain model (Anand) with optimized model parameters for SAC405 solder joints, as described in Sect. 4.5, is used to represent the response of solder joints to applied cyclic flexural loading. The constitutive model is able to account for the strain rate-dependent response of the solder joint. Material model parameters for the Pb-free SAC405 solder joint is as

Table 6.3 Elastic properties of various materials used in the microelectronic assembly model

Material	*E* (GPa)	*G* (GPa)	*v*
SAC405 solder	41.05	–	0.36
Cu layer	117	–	0.34
Nickel layer	200	–	0.31
$(Cu,Ni)_6Sn_5$	85.6	–	0.31
Silicon die	130	–	0.28
Molding compound/ epoxy	25.845	–	0.35
FR-4 PCB and substrate (isotropic plane *x*–*y*)	22 (*x* and *y*) 10 (*z*)	3.9 (*x*–*y*) 3.9 (*x*–*z* and *y*–*z*)	0.11 (*x*–*y*) 0.28 (*x*–*z* and *y*–*z*)

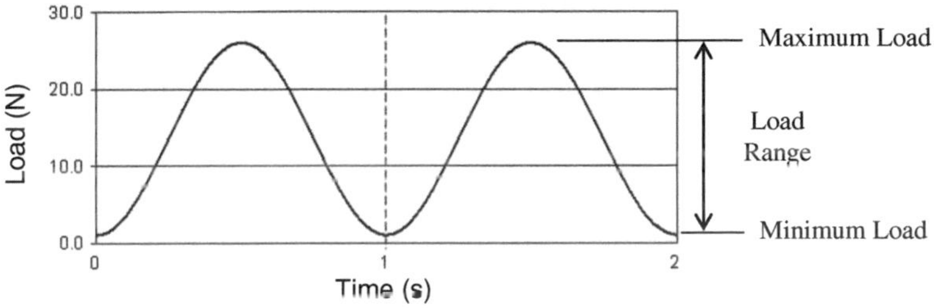

Load Case ID	Load Range ΔP (N)	Load Ratio *R*
LC1	25	0.0385
LC2	30	0.0323
LC3	40	0.0244
LC4	50	0.0196
LC5	60	0.0164

Fig. 6.9 Sinusoidal applied load to the loading rollers for the different flexural load cases

shown in Table 4.5 (Chap. 4). Other materials are assumed to behave elastically with properties gathered from the literature and summarized in Table 6.3. These input properties are similar to those shown in Table 5.1, but computed at the test temperature of 25 °C. The small difference is based on the quoted values from different sources. The various components of the test assembly initially bonded are assumed to remain intact throughout the applied mechanical load cycles.

One of the simulated load cases consisting of controlled sinusoidal force to the loading rollers is illustrated in Fig. 6.9. The positive minimum load of 1 kN will ensure continuous contact between the specimen and the rigid rollers throughout the fatigue cycles. Five different load cases are examined, each with a different load range and a load ratio of minimum to maximum, *R*, as summarized in the accompanied table. The loading frequency is 1 Hz. In this four-point bend test configuration, the PBGA assembly is subjected to a constant magnitude of bending moment.

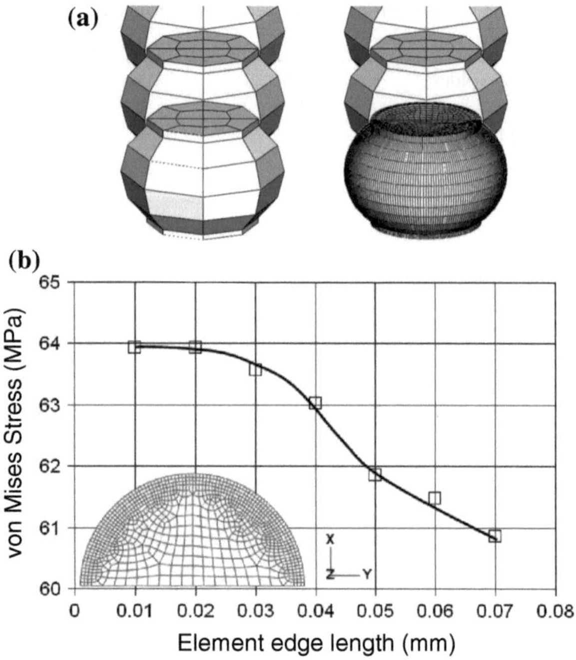

Fig. 6.10 **a** Critical solder joint with refined element mesh, **b** convergence for element size in the selected region of solder/IMC interface in the critical solder joint. The inset figure shows the relative density of the element mesh for the solder

A preliminary run of the FE model with a coarse mesh of 72 elements for each solder joint is performed to identify the critical solder joint in the array. The corner solder joint is found to be the most susceptible to failure based on the predicted inelastic strain. A mesh convergence study is then carried out, as discussed in Sect. 3.3, resulting in the critical solder joint being discretized into 16700 elements, as shown in Fig. 6.10a. In view of the expected local stress gradient in the edge region of the critical solder joint near the solder/IMC interface, a mesh sensitivity analysis is also performed. In this analysis, the edge region of the solder/IMC interface plane is discretized successively into smaller element sizes, while von Mises stress is monitored at a selected material point in the critical region. The resulting variation of von Mises stress with element size is illustrated in Fig. 6.10b. The resulting element mesh on this interface with a maximum edge length of 0.01 mm will ensure that the calculated variables are independent of the size of the elements. In addition, four layers of elements at the solder/IMC interface are meshed over the thickness of 25 μm for use in volume averaging of fatigue-related quantities for use in developing fatigue models of the solder joint.

A previous study indicated that the evolution of accumulated inelastic hysteresis energy per fatigue cycle, $W_{in,acc}$, reaches saturation after four fatigue cycles for Sn-Ag-Cu solder joints [29]. Fatigue-related parameters extracted from such stabilized hysteresis loop are employed in phenomenological-based fatigue models of solder joints. In this respect, only five mechanical load cycles are simulated for each load case described in this chapter.

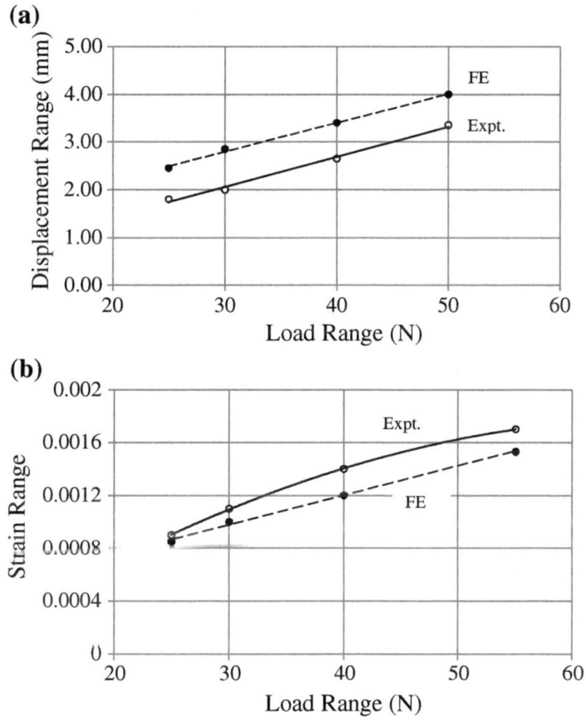

Fig. 6.11 Comparison of FE calculated and measured variables for the different load cases; **a** displacement of loading rollers and **b** strain on surface of the PCB versus applied load range

6.3.2 Deformation of the BGA Test Assembly

The calculated deflection response of the test assembly represented by the displacement range of the loading rollers, to the applied load range is shown in Fig. 6.11a by the dashed line. A measured response of similar quantity is extracted from Kim and Lee [13], and is shown in the same figure for comparison. A relatively linear variation is obtained. The FE predicted displacement ranges are 20.6–30.5 pct. higher than those of measured values. Such discrepancy could easily be introduced by the uncertainty in input properties of the assumed homogeneous orthotropic FR-4 PCB. However, the compliance of the system indicated by the slope of the curve is similar. Thus, the observed discrepancy does not invalidate the deformation mechanics of the assembly under the prescribed loading conditions.

FE calculated surface strain of the PCB is compared with strain gage readings mounted at an identified location on the board, as shown in Fig. 6.11b. The response strain range of the board varies with slight nonlinearity at low applied load range, possibly due to smaller deformation of the board when a large number of solder joints remain elastic. The calculated strain range is lower than measured values for the PCB, although similar correlations are demonstrated. While strain gage measures the true local strains on the surface, the FE calculated value represents an effective strain of the assumed homogeneous orthotropic PCB.

6.3.3 Evolution of Internal States in the Solder Joint

The solder joint located under the corner of the substrate (as indicated by the dark solder joint in Fig. 6.8b) is the most critical solder joint with respect to the accumulated equivalent inelastic strain. Solder joints at an identical location have been observed to fail first under similar flexural load cycles [12]. The largest relative deflection between the substrate and the PCB occurred in the vicinity of the substrate corner region and contributed to the criticality of the solder joints in this region. Typical distribution of von Mises stress and the corresponding equivalent of inelastic strain on the critical solder joint is shown iin Fig. 6.12a, b respectively. The stress and strain fields correspond to the peak of the first load cycle for the load case LC1 ($\Delta P = 25$ N, $P_{max} = 26$ N). It is noted that the stress and inelastic strain fields have similar characteristics to those fields found in the critical solder of the BGA assembly subjected to temperature loading, as discussed earlier (see Fig. 3.11). A critical location with a severe stress gradient and high inelastic strain is confined in the edge region near the solder/IMC interface on the board side of the assembly. The bulk solder joint remains elastic. An identical location of failure was also observed experimentally [12]. In addition, a fatigue crack was observed initiating at this location with extensive plastic strain and propagate preferentially either across the bulk solder or along the solder/IMC interface plane [22]. In view of the similar deformation and failure characteristics, mechanical load cycles can be utilized to quickly generate reliable data if the flexural test can be conducted at a suitable isothermal temperature to ensure an identical mechanism of failure is reproduced as in the case of temperature cycling.

The evolution of von Mises stress and equivalent inelastic strain at the critical material point in the solder (location marked *A* in Fig. 6.12) during the first three bending cycles are shown in Fig. 6.13. The first half of the load cycle to the peak level brings the material point into the plastic zone with the corresponding stress and inelastic strain of 68 MPa and 0.082, respectively. Elastic unloading follows, but only to a relatively low applied load level (about 11 N). It is noted that further unloading to the minimum load level of 1 N causes an additional accumulation of inelastic strain. This is likely due to the effect of local relative deflection between the stiff PBGA package and relatively compliant PCB causing the equivalent von Mises stress to increase. An analysis of each Cartesian stress component would provide a better characteristic evolution of the dominating component. The beginning of reloading in the next cycle causes the internal stress to decrease, thus no increments in inelastic strain occurs. Further loading to the peak load level continuously increases inelastic strain at the material point. This characteristic evolution of von Mises stress and equivalent inelastic strain repeat for subsequent load cycles. Significant inelastic strain (0.101) is induced only during the first bending cycle. A relatively constant smaller magnitude of inelastic strain is accumulated in each subsequent load cycle (0.047 and 0.045 for the second and third cycle, respectively). This validates the earlier assumption of the response of the critical solder in reaching a stabilized stress-strain hysteresis loop after four fatigue cycles.

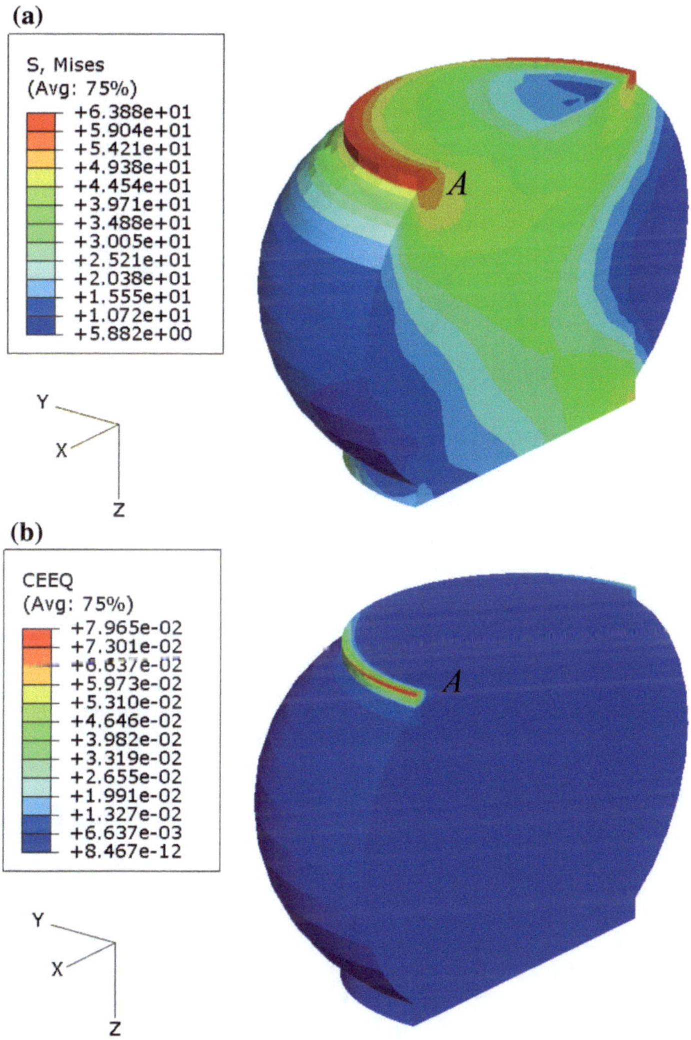

Fig. 6.12 **a** von Mises stress and **b** equivalent inelastic strain distribution in the critical solder joint corresponding to the peak of the first load cycle for load case LC1 ($\Delta P = 25$ N, $P_{max} = 26$ N). Stress scale is in MPa

The critical material point in the solder joint will also experience stress and strain hysteresis during bending cycles leading to extensive accumulation of cyclic inelastic strains. Hysteresis loops for all six Cartesian stress-strain components during the first three bending cycles are shown in Fig. 6.14. The stress-strain response to flexural loading begins from the origin of the plot because thermally-induced residual strain during reflow cooling is not being modeled. Although the bending load fluctuates with a positive load ratio ($R > 0$), a hysteresis loop for

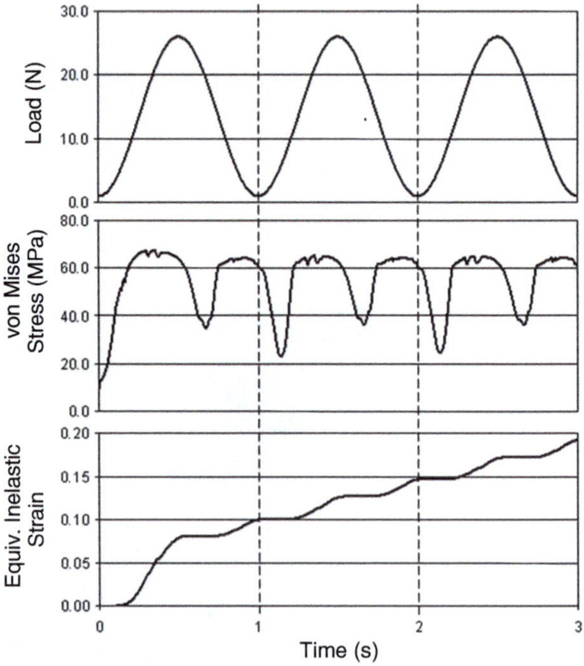

Fig. 6.13 Evolution of von Mises stress and equivalent inelastic strain in the critical solder joint during the first three bending cycles. Loading frequency is 1 Hz

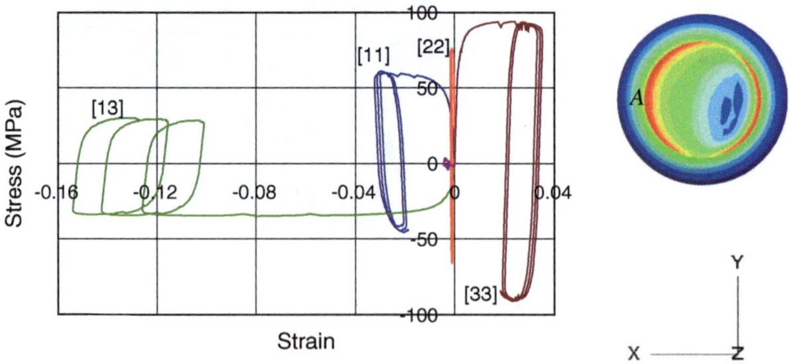

Fig. 6.14 Stress-strain hysteresis loops for the six Cartesian components at the critical material point (marked *A*) in the solder joint. Load case LC1 ($\Delta P = 25$ N, $P_{max} = 26$ N)

each stress component alternates in the stress axis ($R < 0$). This suggests that the critical solder joint experiences reversible bending under the externally applied load cycles. It is noted that the shear strain component, τ_{13}, displays the largest strain range. This localized shear effect is believed to be the main contributor to solder joint fatigue. Such a postulate was employed in modified Coffin-Manson equations for a reliability fatigue model [25]. The largest stress range corresponds

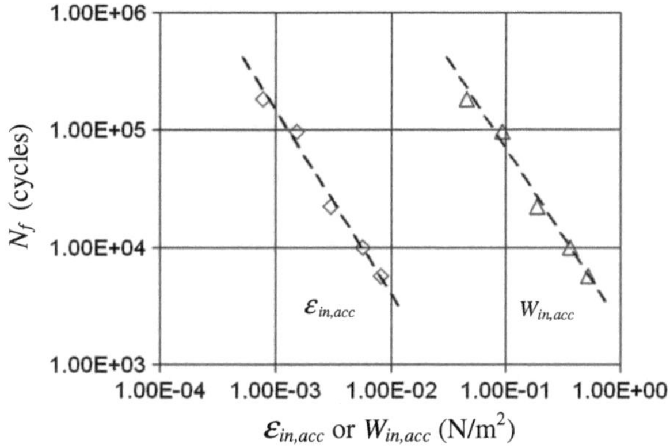

Fig. 6.15 Fatigue life models for cyclic flexural loading of PBGA assembly with SAC405 solder joints

to the normal stress component, σ_{33}, acting on the solder/IMC interface. Mode-I crack opening of the interface is facilitated by the tensile part of this stress cycle. The shear components τ_{12} and τ_{23} are negligible.

6.3.4 Solder Joint Fatigue Life Models

Similar phenomenological-based models as deliberated in Sect. 6.2 are employed to establish fatigue life models for SAC405 solder joints in BGA assembly under cyclic flexural loading. It is noted that the stress-strain hysteresis loops saturate after enduring four load cycles. Fatigue-related variables, namely the accumulated inelastic strain per cycle, $\varepsilon_{in,acc}$, and inelastic work density per load cycle, $W_{in,acc}$, are extracted from predicted data after the last (5th) simulated load cycle. These variables are correlated with a mean fatigue life, N_f, of the PBGA assembly for each load case. The measured mean fatigue life of these assemblies is based on published experimental data under identical loading conditions [12]. The resulting correlation is shown in Fig. 6.15. Fatigue life models for the solder joint under the applied cyclic flexural loading are based on least-squared fit curves of the data, expressed as:
Inelastic strain-based model:

$$N_f = (3.4954\ \varepsilon_{in,acc})^{-1.537} \tag{6.10}$$

Inelastic work density-based model:

$$N_f = (2157.74\ W_{in,acc})^{-1.485} \tag{6.11}$$

As detailed in Sect. 6.2, a fatigue life model with an exponent value of -1 represents fatigue failure by the accumulated creep strains during reliability temperature cycles. An exponent value of approximately -1.5 is obtained for fatigue models of PBGA assembly under cyclic flexural loading. This exponent value is

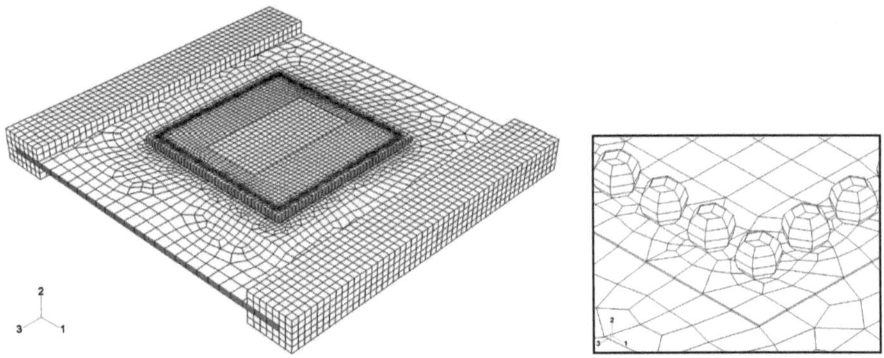

Fig. 6.16 Full (global) model of BGA assembly with end grips used in FE simulation with applied twisting load. Details of the coarse mesh of the peripheral array solder joints (*Right*)

attributed to the mechanically induced strains due to flexural load cycles at relatively low temperatures.

6.4 BGA Assembly Under Cyclic Torsional Load

This chapter describes a reliability simulation of a test assembly with BGA solder joints subjected to cyclic mechanical twisting. Additional requirements in the FE modeling of the assembly are presented. The deformation response of the assembly and distribution of stresses and inelastic strains in the critical solder joint are discussed. The evolution characteristic of inelastic strains in the solder joint during cyclic twisting load (MT1) is compared with those generated by temperature cycles (TC1) and accelerated temperature cycles with a dwell time period (TD1). Fatigue life models based on the inelastic shear strain range and accumulated inelastic strain energy density (or inelastic work density) per load cycle, computed for the critical solder joint, are discussed.

6.4.1 Finite Element Modeling

A full (global) model of the test assembly with BGA solder joints surface mounted on the PCB is created, as shown in Fig. 6.16. The full assembly model is used to establish the overall deformation field and identify critical solder joints in the array. Rigid grip ends are added to receive the applied twisting deformation at one end and fixed displacement boundary conditions at the other. A relatively coarse mesh, particularly in solder joints, is adequate to identify the critical solder joint and establish the global deformation of the test assembly. The BGA test assembly with a peripheral array of Sn-40Pb solder joints is identical to that described in Sect. 3.2, and a quarter model of the assembly is illustrated in Fig. 3.2. The nonlinear viscoplastic

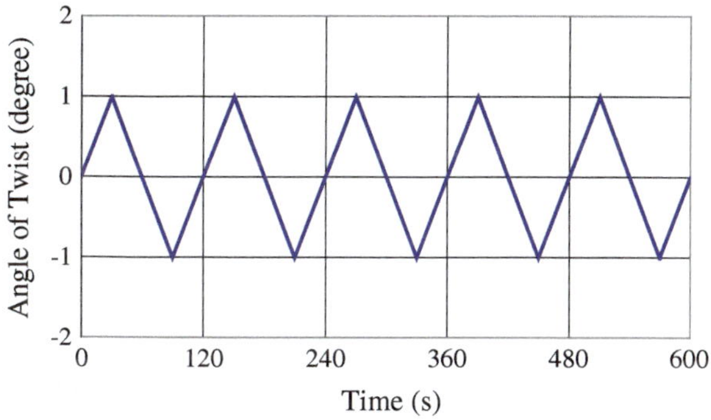

Fig. 6.17 Cyclic twisting profile, MT1 prescribed on the BGA test assembly

response of the Sn-40Pb solder joints is represented by the unified inelastic strain model (Anand model) with model parameter values as listed in Table 3.3. Properties of other materials of the assembly are summarized in Table 3.1. The thin copper traces (0.025 mm-thick foil) are modeled with one layer of linear incompatible mode elements to accurately reproduce the flexing behavior of thin foil.

The twisting load is applied to the assembly by rotating the rigid grip at one end while the other is fixed. The load magnitude is defined by the angle of twist over the length of the assembly between the end grips. The cyclic twisting load profile denoted as MT1 is shown in Fig. 6.17. The completely reversed twisting load has an amplitude of 1° per 50 mm length and a time period of 120 s (loading frequency of 0.00833 Hz). The prior solder reflow cooling process of the assembly is not considered in this simulation.

The predicted deformation response of the full assembly model corresponding to the peak of the first twisting cycle is shown in Fig. 6.18. The displacement contour alternates between the two diagonal symmetry lines of the model at each peak of every twisting cycle. The symmetry of the displacement field about the diagonal planes is demonstrated. Thus, a quarter model of the assembly is adequate for subsequent stress analysis in view of reducing computation time and cost. In addition, the solder joint located at each corner of the peripheral array is identified to be the critical solder joint with respect to plastic strain. This solder joint is discretized into finer element mesh in the sub-modeling process. The mesh convergence study is carried out on the sub-model, as discussed in Sect. 3.3, to ensure that the calculated variables are independent of the size of the elements. The resulting sub-model of the BGA assembly with finer element mesh, particularly for the critical solder joint, is shown in Fig. 6.19. Earlier works have observed that a fatigue crack path preferentially follows the pad-solder interface [5, 24]. Thus, an element layer of 50 μm-thick in the solder at the pad-solder interface is designed for use in averaging the characteristic fatigue variables, namely the inelastic strain energy density per cycle. The calculated displacement field from the global assembly model is imposed on symmetrical planes of the quarter sub-model in the subsequent detailed simulation.

Fatigue responses of the identical BGA assembly with Sn-40Pb solder joints subjected to accelerated temperature cycles (TC1) and accelerated temperature cycles with dwell-time (TD1) are quantified using an FE simulation. The FE

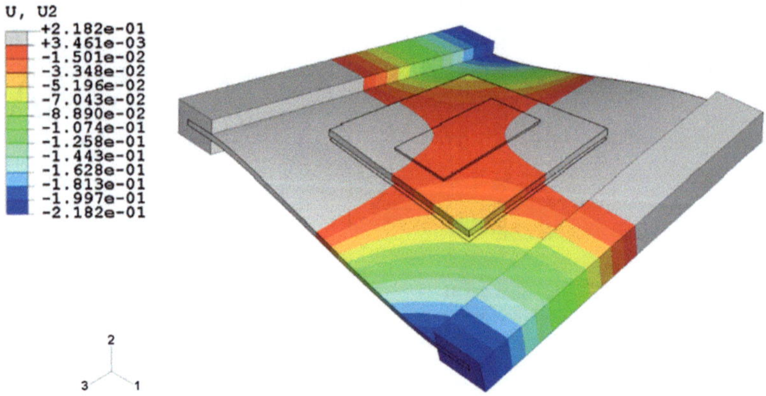

Fig. 6.18 Displacement field (*vertical* component) of the BGA assembly under twisting load. Note that the *gray* region consists of displacements in the positive range up to 0.2182 mm

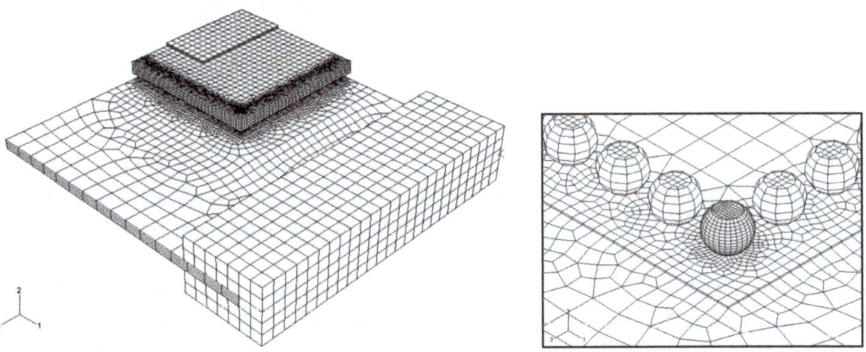

Fig. 6.19 Sub-model of the BGA assembly showing finer element mesh, particularly for the critical corner solder joint. Warpage after solder reflow cooling process. Deformation at the peak of the twisting cycle

modeling procedures are similar to those discussed in Sect. 6.3. The simulated temperature cycles are illustrated in Fig. 3.1b and summarized in Table 6.4. The effects of residual stress and strain induced during the solder reflow process on the subsequent thermal fatigue performance of the solder joints are examined for thermal fatigue cases. The results are to be compared with the assembly subjected to mechanical twisting cycles.

We recall that different locations of the critical solder joint in the BGA assembly are identified when the assembly is subjected to temperature cycles or mechanical load cycles. In the temperature cycling test, the solder joint located at the symmetry plane parallel with the longer side of the Si-die is the most severely strained solder, as illustrated in Fig. 3.3b. However, the solder joint located at the corner of the peripheral array is the most critical in the reliability test with mechanical load cycles. The deformed shape of the assembly, due either to

Table 6.4 Load cases for accelerated temperature and mechanical twisting cycles employed in the simulations

Load case	ΔT (°C)	Ramp rate (°C/min)	t_{dwell} (min)
TC1	−40 to 125 (165)	33	–
TD1	−40 to 125 (165)	33	10
	Degree of twist	Ramp rate (°/s)	t_{dwell} (min)
MT1	1° per 50 mm-length	1/30	–

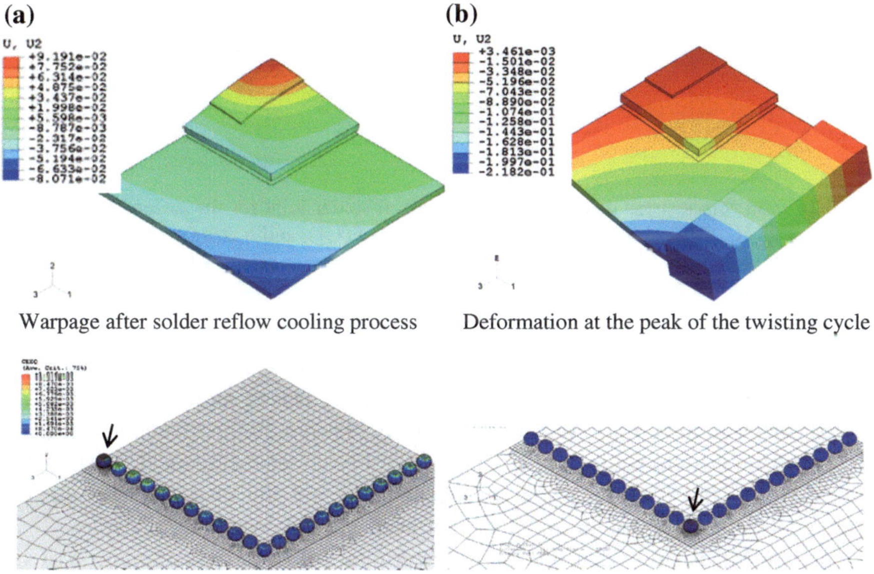

Fig. 6.20 Comparison of the deformed shape of the quarter model assembly and location of the critical solder joint (indicated by the *arrow*) for different types of loading; **a** temperature cycles and **b** mechanical twisting cycles. Note that the displacement scale depends on the current applied load level

temperature or mechanical twisting cycles, induces different relative deformation of the adjacent substrate and PCB, that leads to different locations of the critical solder joint, as compared in Fig. 6.20.

6.4.2 Evolution of Stress and Inelastic Strain in the Critical Solder Joint

Stress and strain fields and their evolution in the critical Sn-40Pb solder joint of the BGA assembly following the solder reflow process and subsequent temperature cycles, identical to TC1, are discussed as an illustrative case study throughout

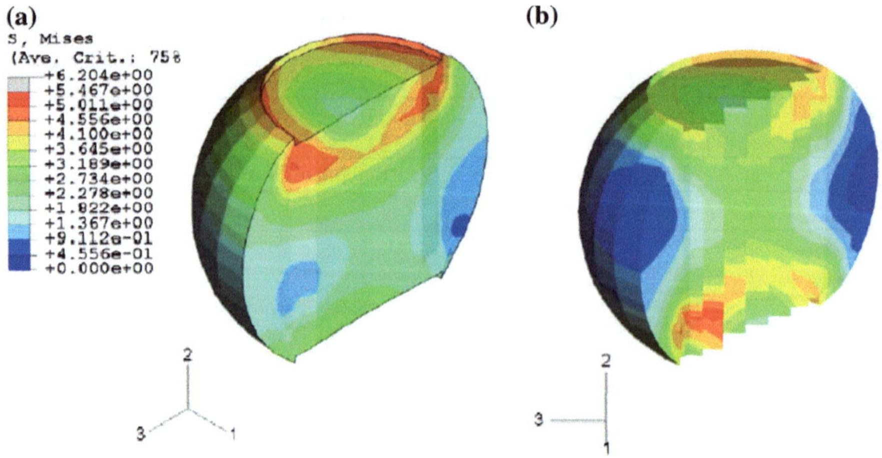

Fig. 6.21 von Mises stress distribution in the critical solder joint for the different fatigue load cases: **a** at 25 °C after completing 4 temperature cycles (TC1), and **b** at +0.5° twist deformation during the 5th cycle (MT1). Note that a smooth cutting plane is used for this figure **a** while the cutting plane for this figure **b** passes along interface boundary of the elements along the path

Chap. 3. The response of the critical solder in a similar BGA assembly under accelerated temperature cycling with a dwell time period similar to TD1 is discussed earlier in Sect. 6.2. Thus, this chapter is focused on the mechanics of deformation of the critical solder in the BGA assembly under mechanical twisting load cycles, MT1, and a comparison of stress fields and the characteristic inelastic strain evolution in the solder under temperature and mechanical load cycles.

The von Mises stress distribution in the critical solder at the completion of specific load cycles is compared in Fig. 6.21. The stress scale is in MPa. The similar characteristic distribution of the stress is predicted for both temperature and mechanical twisting load cases. The highest stress and stress gradient are confined to regions in the vicinity of the solder/IMC interface. However, stress concentration is greatest near the component (top) side for temperature cycling and near the board (bottom) side for cyclic mechanical twisting cycles. Such high stress magnitude is favorable to localized yielding of the solder. A relatively high stress magnitude along the diagonal line on the section plane of the solder is also predicted for both load cases.

The distribution of the accumulated inelastic strain in the critical solder joint at different stages of the load cycles for temperature (TC1) and twisting load (MT1) are compared in Fig. 6.22. The inelastic strain region grows in size with the load cycles, but a small region of highly concentrated strain develops near the edge of the solder/IMC interface. Such highly localized inelastic strain is detrimental in initiating fatigue crack. In the mechanical twisting cycle, the inelastic strain region grows along the diagonal path of the bulk solder where shear strain magnitude is high. This path likely traces the path of crack propagation at a relatively low strain rate. The remaining bulk solder remains elastic. Fatigue crack propagation in the

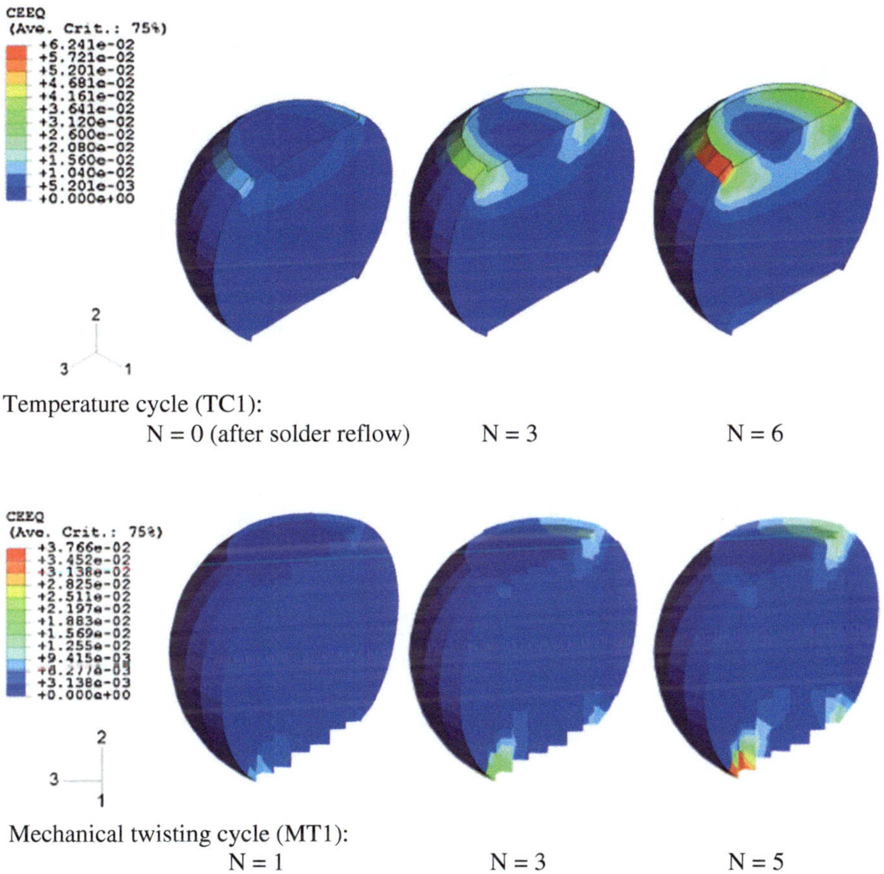

Fig. 6.22 Accumulated inelastic strain fields at the end of the respective load cycles. Note the different cutting plane of symmetry and orientation used for each loading type

bulk solder is dependent on the crack tip driving force or stress intensity factor range and various material factors, including the critical energy release rate of the material. Such fatigue cracking process is discussed in Chaps. 8 and 9 using the damage mechanics concept.

The evolution of equivalent inelastic strain at the selected material point (similar to point *A* in Fig. 6.12) in the critical solder joint is compared in Fig. 6.23 for the three different load cases. Both thermal fatigue cases (TC1 and TD1) started with an initial residual inelastic strain of 0.0137 induced during the solder reflow cooling process. The rate of inelastic strain accumulation per fatigue cycle for temperature cycling (TC1) and mechanical twisting load cycles (MT1) is similar at 0.0075 per cycle, as depicted by the same nominal slope of the curves. Thus, the solder reflow process-induced inelastic strain does not indicate any influence in the predicted fatigue response of the solder joint. The 10-min dwell time period at both peak temperature levels (TD1) accelerates the accumulation of inelastic strain

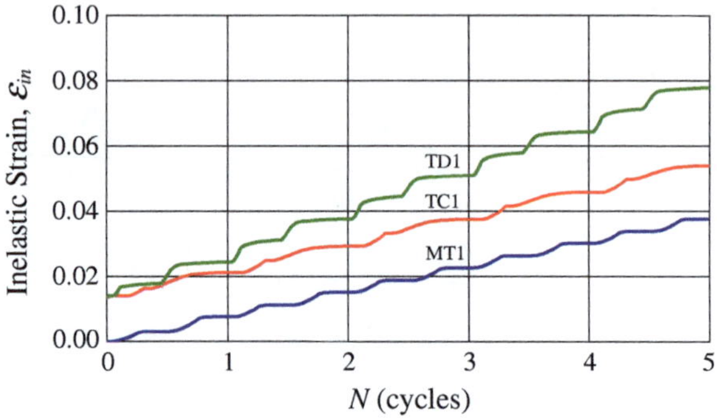

Fig. 6.23 Evolution of equivalent inelastic strain in the critical Sn-40Pb solder joint for the different fatigue load cases

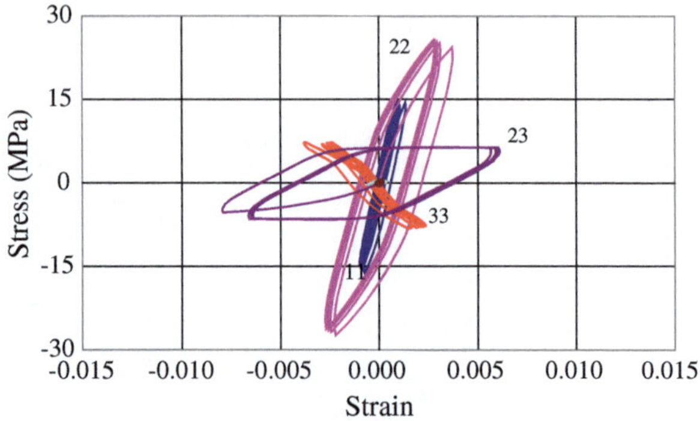

Fig. 6.24 Hysteresis loops for Cartesian stress-strain components during mechanical twisting cycles at 25 °C. Indices 11, 22, 33 and 23 refer to the Cartesian coordinate axes used in the FE model

in the solder joint as reflected in the higher slope of the curve. Stress relaxation that manifests over the hold-time duration of the cycles is accompanied by creep and viscoplastic deformation that leads to faster inelastic strain rate per cycle. Results indicate that combinations of twisting load amplitude and isothermal testing temperature could generate the characteristic curve of inelastic strain evolution corresponding to specific temperature cycling tests. In addition, a mechanical fatigue test of the BGA assembly at an isothermal temperature is expected to generate similar creep-fatigue interaction effects to the temperature cycling test. Thus, mechanical twisting load cycles could replace temperature cycles as a faster and cheaper test method in generating the much-needed solder joint reliability data.

The stress-strain hysteresis response of the solder joint material for the first five mechanical twisting cycles of the BGA assembly is shown in Fig. 6.24. Since the solder reflow process is not considered in this case, each hysteresis loop begins at the origin of the stress-strain axes (The effect of the initial residual stresses and strains on subsequent hysteresis loops is illustrated in Fig. 6.14 for cyclic flexural loading). The shear component (in 2–3 directions, parallel to the solder/pad interface plane) displays the largest total strain range, $\Delta\gamma = 0.012$, with a corresponding stress range of $\Delta\tau = 13.0$ MPa. The large relative shear displacement within the solder/IMC interface intensifies the cyclic interfacial shear stress and strain resulting in crack initiation, as observed experimentally [22]. The largest stress

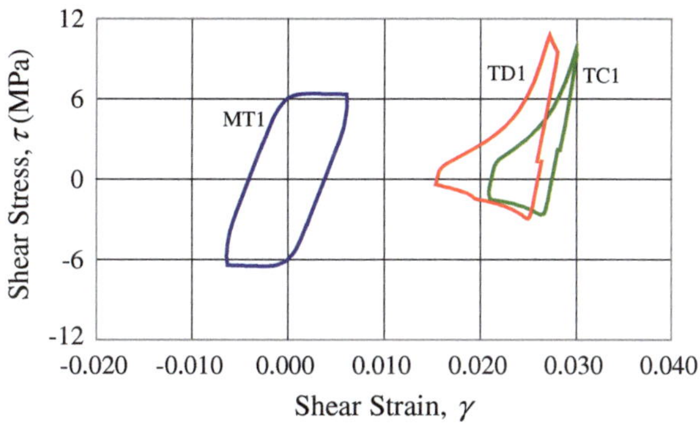

Fig. 6.25 Stabilized shear stress-strain hysteresis loops of the critical material point in the solder joint under accelerated temperature cycle (*TC1*), accelerated temperature cycle with dwell-time (*TD1*) and mechanical twisting load cycles (*MT1*)

range of 50.0 MPa is predicted for the normal component to the interface plane (2–2 direction). The tensile part of this stress is favorable for crack initiation under the opening Mode-I loading at the edge of the solder/IMC interface. It is worth noting that the ranking of stress and strain ranges by magnitude is similar for both cases of temperature cycling and the mechanical twisting load cycles.

Since the shear component displays the largest inelastic strain range, the inelastic shear strain range is postulated to govern the fatigue life of solder joints. The saturated shear stress-strain hysteresis loops represented by the 5th cycle of each fatigue load cases are compared in Fig. 6.25. Hysteresis loops for temperature cycles TC1 and TD1 are shifted away from the coordinate origin due to the locked-in residual stresses and strains following the initial solder reflow cooling process. Although the shape of the shear hysteresis loops is different for thermal and mechanical fatigue (for load case TC1 and MT1), the stress and inelastic strain ranges are comparable. The inelastic shear strain range is extracted from the hysteresis loop at zero stress after the recoverable elastic part is subtracted (similar to that described in Fig. 4.17).

Inelastic shear strain range, $\Delta\gamma_{in}$, and accumulated inelastic work density per cycle, $W_{in,acc}$, were identified as fatigue-related variables for the fatigue life prediction model of solder joints, as discussed in Sect. 4.6 (see Eqs. 4.30–4.33). Parameter values for the strain-based fatigue model has been determined for the Sn-40Pb solder at −50, 35 and 125 °C to yield the following [25]:

$$N_f = 1.2928(\Delta\gamma_{in})^{-1.96} \tag{6.12}$$

Similarly, the energy-based fatigue life model for the solder joint (Eq. 4.31) has been established as [3, 4]:

$$\begin{aligned} N_f &= N_o + \frac{a}{da/dN} \\ &= 13,173(W_{in,acc})^{-1.45} + \frac{a}{3.92(10^{-7})(W_{in,acc})^{1.12}} \end{aligned} \tag{6.13}$$

Table 6.5 Predicted fatigue lives of Sn-40Pb solder joints in BGA assembly using strain-based model, Eq. (6.12) and energy-based model, Eq. (6.13)

Load case	$\Delta\gamma_{in}$	$W_{in,acc}$ (psi)	Predicted life, N_f (cycles [days])	
			Strain-based	Energy-based
			Solomon [25]	Darveaux [4]
TC1	0.00620	1.566	27,556 [191]	34,212 [238]
TD1	0.01026	2.668	10,263 [214]	18,227 [380]
MT1	0.00817	2.088	–	24,337 [34]

The characteristic crack length, a, corresponds to the edge length of the element in the converged mesh. Fatigue-related variables are extracted from the FE simulation results for the different fatigue load cases. These variables are then used to examine the predictive capability of fatigue life models shown in Eqs. (6.12) and (6.13). Results are summarized in Table 6.5.

The predicted fatigue life based on the phenomenological approach is associated with low accuracy in view of the complex nature of solder creep behavior and uncertainty in board level temperature cycling tests [18]. However, the different creep models and finite element assumptions only affect absolute fatigue life, but not the relative trend of fatigue performance [18, 33]. This relative design trend has been evaluated in terms of the accumulated inelastic work density per cycle, $W_{in,acc}$, with a reasonable range of fatigue lives. Previous work on different TFBGA packages with eutectic solders indicated that $W_{in,acc}$ ranges from 0.154 to 0.175 MPa [18]. In an experimental validation study, 2512-type resistors with Sn-40Pb solder joints were subjected to temperature ranges of −55 to 125 °C (4.5 °C/min) and −40 to 120 °C (4.0 °C/min). The calculated $W_{in,acc}$ is 7.3 and 6.0 MPa, respectively [6]. The low magnitude of plastic work density calculated in this simulation ($W_{in,acc}$ = 0.0108 MPa or 1.566 psi for load case TC1) is attributed to the fast temperature ramp rate of 33 °C/min. This fast temperature ramp rate lowers the relative accumulation of inelastic strains per cycle due to a smaller creep effect, thus contributing to longer characteristic life (cycles).

As expected, it is noted that the accelerated temperature cycling with a 10-min dwell time period (TD1) results in a shorter life (smaller number of fatigue cycles to failure) of the solder joint compared to temperature cycling (TC1). However, the corresponding reliability testing time is longer. In this respect, a thermal shock test is perhaps a favorable alternative to an accelerated temperature cycling test. Cyclic mechanical twisting (MT1), for a similar rate of inelastic strain accumulation, predicts a moderate number of fatigue cycles to failure, but with a much shorter testing time. An isothermal mechanical test at an elevated temperature could further accelerate the fatigue damage process. Correlations between cyclic mechanical twisting data and temperature cycling test data can be made using Weibull statistics to obtain the acceleration or scale factor for the given geometry and design of the assembly. In addition, isothermal test conditions and mechanical twisting load magnitude and frequency are easily achieved in an actual reliability test. In this respect, FE simulations can be effectively utilized to optimize the accelerated cyclic mechanical test profile.

References

1. Amagai, M.: Chip scale package (CSP) solder joint reliability and modeling. Microelectron. Reliab. **39**, 463–477 (1999)
2. Che, F.X., Yap, D., Luan, J.E., Goh, K.Y., Ma, Y.Y.: Design for cyclic bending reliability of large PBGA assembly using experimental and numerical methods. In: Proceedings of 2008 Electronics Packaging Technology Conference, pp. 1171–1177 (2008)
3. Darveaux, R., Mawer, A.: Thermal and power cycling limits of plastic ball grid array (PBGA) assemblies. In: Proceeding of Surface Mount Technology, pp. 315–326 (1995)
4. Darveaux, R.: Solder joint fatigue model. In: Proceeding of TMS Conference (1997)
5. Dudek, R., Nylen, M., Schubert, A., Michel, B., Reichl, H.: An efficient approach to predict solder fatigue life and its application to SM and area array components. In: Proceedings of the 1997 Electronic Components and Technology Conference, pp. 462–471 (1997)
6. Dusek, M., Nottay, J., Hunt, C., Lu, H., Bailey, C.: An Experimental Validation of Modeling for Pb-Free Solder Joint Reliability. National Physical Laboratory Report, NPL MATC (A) 11, Middlesex, UK, Oct (2001)
7. Geng, P., Chen, P., Ling, Y.: Effect of strain rate on solder joint failure under mechanical load. In: Proceedings of 2002 Electronic Components and Technology Conference, pp. 974–978 (2002)
8. Harada, K., Baba, S., Wu, Q., Matsushima, H., Matsunaga, T., Uegai, Y., Kimura, M.: Analysis of solder joint fracture under mechanical bending test. In: Proceedings of 2003 Electronic Components and Technology Conference, pp. 1731–1737 (2003)
9. IPC/JEDEC Standard IPC/JEDEC-9702: Monotonic bend characterization of board-level interconnects (2004)
10. JEDEC Standard JESD22B113: Board level cyclic bend test method for interconnect reliability characterization of components for handheld electronic products (2006)
11. JESD22-A104B: Temperature Cycling, Electronic Industries Alliance/Joint Electron Device Engineering Council (EIA/JEDEC) (2000)
12. Kim, I., Lee, S.-B.: Reliability and failure analysis of lead-free solder joints for PBGA package under a cyclic bending load. IEEE Trans. Compon. Packag. Technol. **31**, 478–484 (2008)
13. Kim, I., Lee, S.-B.: Reliability assessment of BGA solder joints under cyclic bending loads. In: Proceedings of 2005 International Symposium on Electronics Materials and Packaging, pp. 27–32 (2005)
14. Lee, T., Lee, J., Jung, I.: Finite element analysis for solder ball failures in chip scale package. Microelectron. Reliab. **38**(12), 1941–1947 (1998)
15. Liew, Y.B.: Failure assessment of surface mount assembly under thermo-mechanical loading conditions. Master's thesis, Universiti Teknologi Malaysia, Malaysia (2006)
16. Manson, S.S.: Thermal Stress and Low Cycle Fatigue. McGraw-Hill, New York (1966)
17. Mercado, L.L., Phillips, B., Sahasrabudhe, S., Sedillo, J.P., Bray, D., Monroe, E.: Use-condition-based cyclic bend test development for handheld components. In: Proceedings of 2004 Electronic Components and Technology Conference, pp. 1279–1287 (2004)
18. Ng, H.S., Tee, T.Y., Goh, K.Y., Luan, J.-E., Reinikainen, T., Hussa, E., Kujala, A.: Absolute and relative fatigue life prediction methodology for virtual qualification and design enhancement of lead-free BGA. In: Proceedings of the 2005 Electronic Components and Technology, pp. 1282–1291 (2005)
19. Pang, H.L.J., Ang, K.H., Shi, X.Q., Wang, Z.P.: Methodology for highly accelerated solder joint reliability test. In: Proceedings on Electronic Packaging Technology Conference, pp. 385–390 (2000)
20. Pang, H.L.J., Che, F.-X.: Isothermal cyclic bend fatigue test method for lead free solder joints. In: Proceedings of Intersociety Conference on Thermal and Thermo mechanical Phenomena in Electronics Systems, pp. 1011–1017 (2006)
21. Pang, H.L.J.: Sensitivity study of temperature and strain rate dependent properties on solder joint fatigue life. In: Proceedings of IEEE Electronic Packaging Technology Conference, pp. 184–189 (1998)

22. Rooney, D.T., Castello, N.T., Cibulsky, M., Abbott, D., Xie, D.: Materials characterization of the effect of mechanical bending on area array package interconnects. Microelectron. Reliab. **44**, 275–285 (2004)
23. Schubert, A., Dudek, R., Auerswald, E., Gollhardt, A., Michel, B., Reichl, H.: Fatigue life models for Sn-Ag-Cu and Sn-Pb solder joints evaluated by experiments and simulation. In: Proceedings of 53rd Electronic Components and Technology Conference, pp. 603–610 (2003)
24. Shetty, S., Reinikainen, T.: Three- and four-point bend testing for electronic packages. ASME Trans. J. Electron. Packag. **125**, 556–561 (2003)
25. Solomon, H.D.: Fatigue of 60/40 solder. IEEE Trans. Compon. Hybrids Manufact. Technol. **9**(4), 423–432 (1986)
26. Stout, E.A., Sottos, N.R., Skipor, A.F.: Mechanical characterization of plastic ball grid array package flexure using Moiré interferometry. IEEE Trans. Adv. Packag. **23**, 637–645 (2000)
27. Syed, A.: Updated life prediction models for solder joints with removal of modeling assumptions and effect of constitutive equations. In: Proceedings of the 2006 International Conference on Thermal, Mechanical and Multi-Physics Simulation Experiments in Microelectronics and Micro-Systems (EuroSimE), pp. 1–9 (2006)
28. Tamin, M.N., Liew, Y.B., Wagiman, A.N.R., Loh, W.K.: Solder joint fatigue in a surface mount assembly subjected to mechanical loading. IEEE Trans. Compon. Packag. Technol. **30**(4), 824–829 (2007)
29. Tunga, K.R.: Study of Sn-Ag-Cu alloy reliability through material microstructure evolution and laser moire interferometry. Ph.D. thesis, School of Mechanical Engineering, Georgia Institute of Technology, USA (2008)
30. Wiese, S., Meusel, E., Wolter, K.-J.: Microstructural dependence of constitutive properties of eutextic Sn-Ag and Sn-Ag-Cu solders. In: Proceedings of Electronic Components and Technology Conference, pp. 197–206 (2003)
31. Wu, J.D., Ho, S.H., Zheng, P.J., Liao, C.C., Hung, S.C.: An experimental study of failure and fatigue life of a stacked CSP subjected to cyclic bending. In: Proceedings of 2001 Electronic Components and Technology Conference, pp. 1081–1086 (2001)
32. Yeo, A., Lee, C., Pang, H.L.J.: Flip chip solder joint fatigue life model investigation. In: Proceedings of Electronics Packaging Technology Conference, pp. 107–114 (2002)
33. Zahn, B.A.: Solder joint fatigue life model methodology for 63Sn37Pb and 95.5Sn4Ag0.5Cu materials. In: Proceedings of Electronics Components and Technology Conference, pp. 83–94 (2003)
34. Zhai, C.J., Sidharth, Blish, R., II.: Board level solder reliability versus ramp rate and dwell time during temperature cycling. IEEE Trans. Devices Mater. Reliab. **3**(4), 207–212 (2003)

Chapter 7
Damage Mechanics-Based Models

7.1 Introduction

The mechanics of materials is a branch of mechanics that investigates the response of engineering materials and structures to loading and environment. It relates the externally applied load to internal states of the material, namely displacement, strain and stress, and their dependency on temperature and strain rate. The deformation and failure process of a material is a complex process involving nonlinear behavior and different mechanisms. In this respect, a description of the mechanics of materials is separated into two primary domains, namely continuum mechanics and fracture mechanics, as illustrated in Fig. 7.1. The simulation of materials response using constitutive and failure models developed in each domain is limited by an assumption on the nature of the continuum and the length scale used.

Within the continuum mechanics domain, the response of materials and structures to loading and environment is described with the assumption that the material behaves as a continuous medium. Constitutive models represent the unique stress-strain relationship of the material, covering elastic deformation, yielding and hardening behavior. Continuum mechanics analysis ceases to be valid once crack is initiated, thus a prediction of fracture is not possible. Fundamental theories describing the mechanics of materials within this domain are described in Chap. 4. In the fracture mechanics domain, a structural crack is assumed to have existed in the material. Therefore, linear elastic fracture mechanics analysis is aimed at establishing the relationship between the applied load, crack geometry and fracture toughness of the material. The crack tip stress field is quantified using a stress intensity factor and used with fracture toughness to design against a catastrophic fracture of the material. A fatigue crack growth analysis establishes the threshold stress intensity factor range below which the pre-existing crack will not grow. If stable crack propagation occurs, fatigue life can be predicted based on the crack growth rate of the material. FE simulation in the fracture mechanics domain

M. N. Tamin and N. M. Shaffiar, *Solder Joint Reliability Assessment*,
Advanced Structured Materials 37, DOI: 10.1007/978-3-319-00092-3_7,

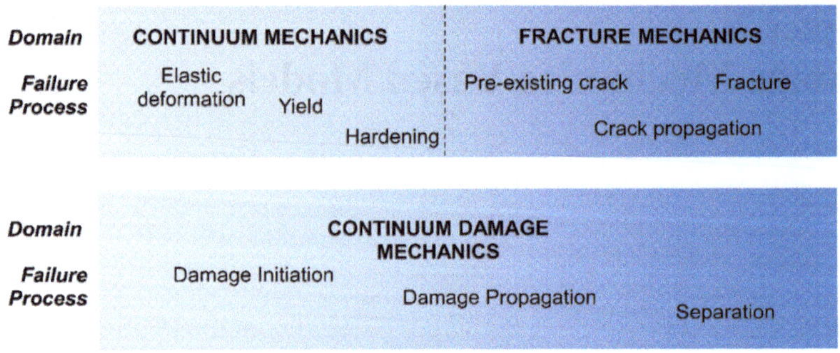

Fig. 7.1 Analysis domains for the mechanics of solids

for solder joint reliability prediction is not covered in the scope of this book. Readers may find useful information on the fundamentals of fracture mechanics and analysis in numerous publications (e.g. [1, 4]).

A description of the continuous process of material failure from a reference pristine condition to the final fracture is made possible with analysis in the continuum damage mechanics domain (see Fig. 7.1). The theory is used to analyze damage and fracture processes in materials at different length scales of the continuum. The failure process is described as successive stages of damage initiation and propagation with the accompanying deterioration in mechanical properties and final separation of the material. Therefore, the tasks in damage mechanics analysis include the followings:

- Determining conditions for the onset/initiation of the first damage event in the material.
- Characterizing and quantifying damage and its evolution following the initiation event.
- Establishing the effect of damage on the materials response.
- Defining the criticality of the damage for assessment of the performance of the material.
- Incorporating the damage concept in structural analysis.

This chapter reviews the material damage concept within the continuum mechanics framework and describes the various length scales for damage analysis. In-depth theory and analysis of continuum damage mechanics is found in open literature (e.g. [28]). The two approaches of damage modeling are employed, namely the continuum and the local approach. The Continuum Damage Model (CDM) is described for the prediction of failure of the bulk solder joint, while the later utilizes a Cohesive Zone Model (CZM) for the failure of thin solder/IMC interfaces. The extension of these models to account for the fatigue performance of materials is adequately addressed. An FE simulation of solder joint reliability employing these damage models is illustrated.

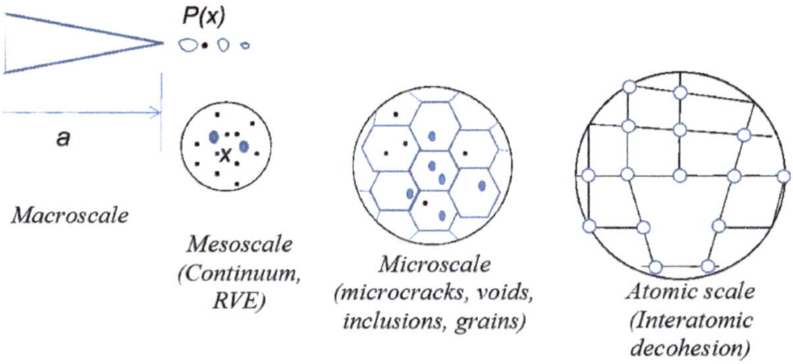

Fig. 7.2 The different length scales of damage observation

7.2 Material Damage Concept

Damage is the development of surface or volume-like discontinuities in a material at various length scales. Damage can be quantified in terms of inherent microstructure defect, flaws due to the processing of the materials, or those accumulated throughout the loading history of the component. Damage is treated as a material internal state variable that initiates and subsequently evolves upon loading, thus causing degradation of the strength properties of the material. It is closely linked to the failure process of the material, while fracture represents the final separation event of the adjacent material points. In a simulation of the material failure process, damage variables are quantified in terms of physical parameters, such as plastic strains, micro-cracks and porosity, or defined using a mechanics parameter, expressed in terms of stress tensor or plastic work density at the material point in question. Processes involving damage phenomena include ductile damage, creep damage in metals, matrix cracking, interface delamination and fiber fracture in composites, and corrosion fatigue. Damage evolution is an irreversible process of energy dissipation in a failing material.

The length scales or range of observation used in the analysis determines the various aspects of the damage phenomena, as illustrated in Fig. 7.2. A crack in a sufficiently large plate could represent material damage on the macroscale. In this scale, material point is assumed distributed continuously in the material. If a mechanical property or state averaged over a small region of a representative volume element with a material point $P(x)$ can be expressed as a continuous function of the spatial position in the volume, then the material can be idealized as a continuum in the mesoscale. Analysis in the microscale accounts for discontinuous structures, such as microcavities and microcracks, in the material. Damage of materials at the atomic scale represents decohesion induced by the separation of interatomic bonds.

7.2.1 Mechanical Representation of Damage

The damage state in a material is represented in terms of scalar damage variables, D ($0 \leq D \leq 1$), where the value of $D = 0$ represents the initial undamaged state and $D = 1$ signifies the final fractured state of the material. The model describes the evolution of degradation phenomena on the microscale from the initial (undamaged/pre-damaged) state up to the creation of a crack on the mesoscale (material element) level. The damage initiation event and subsequent damage propagation process to final fracture of a structure is described using specific criterion and a pre-defined material softening law, respectively, as monitored at material points throughout the structure. Damage is described using an effective stress and mechanical equivalence concept between damaged and undamaged materials. The development of damage is expressed using an evolution equation defined as [20, 31]:

$$\dot{D} = A\left(\frac{\sigma}{1-D}\right)^m \tag{7.1}$$

where ($\bullet$) and σ denote time derivative and stress, respectively, while A and m are material constants. The mechanical representation of the damage state in a material is discussed here based on the effective area reduction due to voids development. The effective area, $\widetilde{dA}$, that carries internal force, dF, on the surface element, dA, containing voids with total area, $dA_{D,}$ can then be defined as:

$$\widetilde{dA} = dA - dA_D \tag{7.2}$$

The damage variable can be interpreted as the fraction of the decrease in the effective area due to void development:

$$D = \frac{dA_D}{dA} \tag{7.3}$$

Combining Eqs. (7.3) and (7.2) yield:

$$\widetilde{dA} = (1-D)dA \tag{7.4}$$

Subsequently, the stress induced by the external force, dF, is increased due to the decrease in the load-carrying area. This effective stress, $\widetilde{\sigma}$ is given by:

$$\widetilde{\sigma} = \frac{dF}{\widetilde{dA}} = \frac{dF}{(1-D)dA} = \frac{\sigma}{1-D} \tag{7.5}$$

Thus, a damaged material subjected to a stress, σ is mechanically equivalent to a fictitious undamaged material subjected to an effective stress, $\widetilde{\sigma}$. The effective stress does not only represent the effect of the area reduction due to voids development, but also an effect of stress concentration at the voids and interactions among these voids. This concept of mechanical equivalence of undamaged and damaged states of a material is illustrated in Fig. 7.3. The damage state can also be

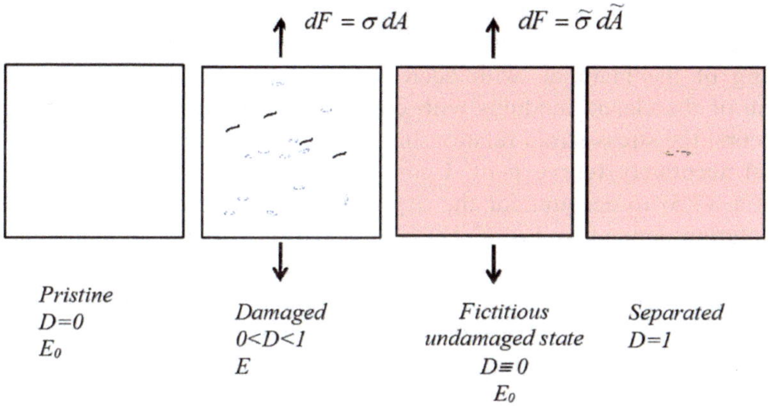

Fig. 7.3 Concept for mechanical equivalence of damage at mesoscale

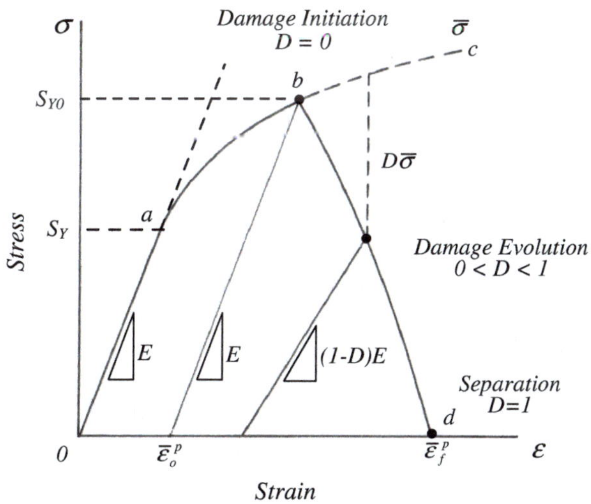

Fig. 7.4 Concept for continuum damage mechanics of a material point undergoing the failure process of damage initiation event, damage evolution and final separation of the material point

characterized in a similar manner by the variation in elastic modulus [6, 25] and void volume fraction [16, 33]. A comprehensive treatment of continuum damage mechanics is available elsewhere (e.g. [28]).

The response of damaged material to loading can be visualized using a uniaxial true stress-strain curve, as illustrated in Fig. 7.4. Hooke's law describes the initial elastic deformation of a damage-free material up to the initial yield (point *a*) with the corresponding initial yield strength, S_Y. Continuous loading beyond the yield point induces plastic strain with an increased yield strength of the hardening material. At sufficiently high stress, voids initiate in the ductile material to signify the onset of damage (point *b*). This damage initiation event corresponds to subsequent yield stress level, S_{Y0}, and equivalent plastic strain, $\bar{\varepsilon}_0^p$.

An explicit criterion is required to identify the onset of damage initiation. In an elastic-plastic material with isotropic hardening, damage manifests in the softening of the material, with decreasing subsequent yield stress and degradation of the elastic modulus with continuous loading. When material damage occurs, the stress-strain relationship ceases to represent the behavior of the material accurately (curve *b–c*). Upon damage initiation, the stress is scaled using Eq. (7.5) to account for the degradation of the modulus with the damage evolution defined by Eq. (7.1), as depicted by curve *b–d*. Simultaneously, the degrading elastic modulus is scaled accordingly. Separation of the material point occurs (point *d*) when the degrading modulus is lost with the corresponding equivalent plastic strain at separation, $\bar{\varepsilon}_f^p$. In a tension test, the fracture event is accompanied by localized necking at the fractured location. Consequently, a three-dimensional stress state prevails and influences the magnitude of plastic strain at fracture.

It is desirable to simulate the fracture process of solder joints in BGA assembly as part of a reliability assessment of the assembly. In reliability testing, the BGA assembly is subjected to the reflow cooling process to form the solder interconnection and subsequently subjected to reliability temperature and/or mechanical cycles. Consequently, the failure of the solder joint is governed primarily by low cycle fatigue and the creep-fatigue interaction mechanism of the solder alloy. It has been observed that crack initiates at the edge of the solder/IMC interface of the critical solder joint under fatigue loading. The fatigue crack is likely to propagate into the bulk solder joint if the local acting strain rate is relatively low. However, under rapid straining of the material point, cracking across the relatively brittle solder/IMC interface offers a favorable option. Such high strain rate loading can be induced in BGA solder joints of the assembly during the reliability board-level drop test. The increasing strength of a ductile solder and decreasing strength of solder/IMC interface with strain rate has been reported for selected reflowed solder/substrate systems [8]. Consequently, damage-based models are required to capture the different fatigue failure mechanisms for the bulk solder and the relatively brittle solder/IMC interface. In this respect, the cohesive zone model is proposed for the prediction of solder/IMC interface fractures, while the continuum damage model is recommended for the simulation of bulk solder failure. In addition, these damage-based models should also account for the observed temperature- and strain rate-dependent failure processes.

7.3 Cohesive Zone Model

The cohesive zone model (CZM) has been proposed and investigated by many researchers, including Dugdale [15], Barenblatt [2] and Rice [30]. The theory postulates fracture as a gradual phenomenon in which separation of materials takes place across an extended *crack tip*, or cohesive zone, and resisted by cohesive traction. The cohesive zone model is discussed here in view of its potential application in reliability assessment of solder joint by simulating the relatively brittle solder/IMC interface. The basic concept of cohesive behavior and key equations used in describing the model is presented. Procedures for the determination of CZM parameter values are elaborated. An extension of the model to account for cyclic loading of the interface is adequately described.

7.3.1 Cohesive Behavior

The cohesive behavior of an interface can be described in terms of an idealized traction-relative displacement response of a double-cantilevered beam with a cohesive interface, as illustrated in Fig. 7.5a. The beam is subjected to an opening (Mode I) load. Two material points, initially positioned on opposite sides of the interface, are considered, and the interfacial traction is taken to depend only on the displacement difference across the interface. The mechanical response of the cohesive interface is assumed, as a first approximation to be represented by a bilinear traction-displacement softening law as shown in Fig. 7.5b. An initial applied load causes a linear increase of traction, σ with (normal) displacement, δ_n denoting elastic behavior, of the material point at the interface. The slope of this line, 0–*a*–*b*, represents the penalty stiffness of the interface, k_n. The traction reaches a maximum value, *T* that represents the tensile strength of the interface. This corresponds to the onset of damage due to excessive normal displacement at point *b*. The corresponding damage initiation displacement is denoted by δ_n^o. Further loading causes damage evolution with simultaneous degradation of the property (penalty stiffness) of the interface, as depicted by line *b*–*c*–*d*. The load-carrying capability of the cohesive interface is reduced, as reflected by lower stress with increasing damage displacement. At point *d* the accumulated damage reaches the critical level for separation of the interface. This condition is governed by the critical energy released rate of the interface, G_{IC} as represented by the area under the traction-displacement curve. The stress is reduced to zero to acknowledge traction-free surfaces of the separated interface elements. A material point marked *e* corresponds to the separated interface material element.

The bilinear curve (Fig. 7.5b) represents the traction-displacement response experienced by the interface material at point *e* (shown in Fig. 7.5a), while other points on the curve indicate traction-displacement conditions of the respective points in Fig. 7.5a. Similar traction-relative displacement behavior is described for shear loading of the interface, as would be found in identical beam (end-notch fracture, ENF) specimens loaded under three-point bending to induce dominant Mode II loading conditions. The fracture-related parameters as shown in Fig. 7.5c are penalty stiffness, k_s, interface shear strength, *S*, shear displacements, $\left(\delta_s^o, \delta_s^f\right)$ and Mode-II critical energy release rate, G_{IIC}. In addition, it is assumed that the response of the cohesive interface under tearing Mode-III loading is identical to that of pure shear.

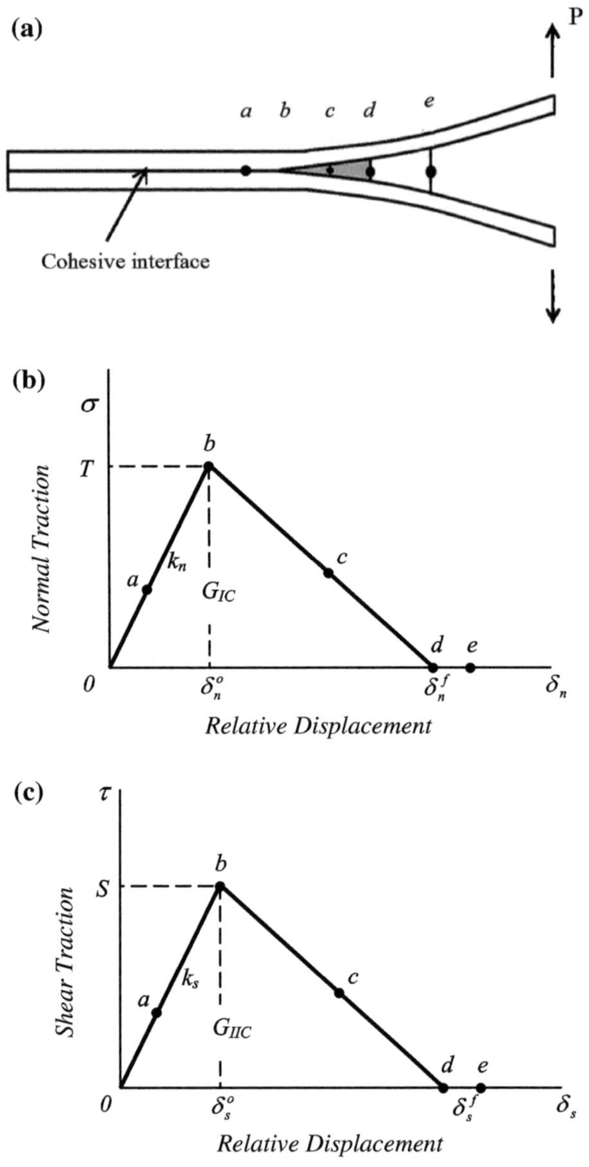

Fig. 7.5 Concept for cohesive behavior of interfaces: **a** interface under opening load, **b** bilinear traction-relative displacement in tensile (Mode I) loading and, **c** shear (Mode II or Mode III) failure of the interface

7.3.2 Key Equations for Cohesive Zone Model

The mechanics of bi-material interfaces have been well studied, particularly in laminated composite materials and structures [22, 38]. The cohesive damage zone concept has been employed to examine the interface cracking process under mixed-mode loading conditions [10, 22]. In applying the model for the solder/IMC interface fracture process in solder joints, a thin interface (with mathematically zero thickness)

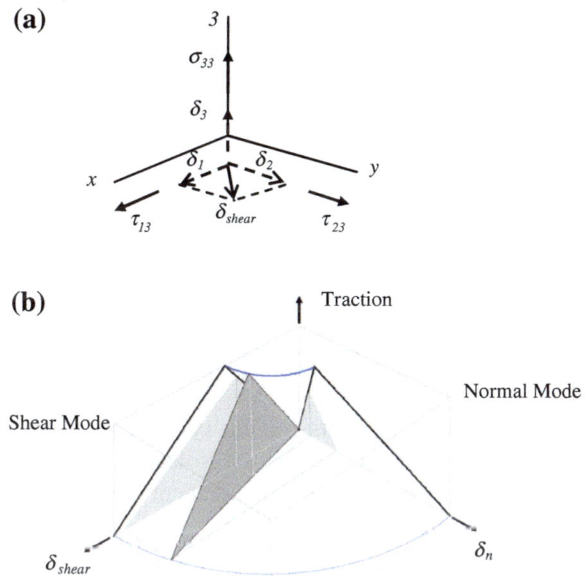

Fig. 7.6 **a** Traction and displacement components acting on interface plane *x–y*. **b** Traction-relative displacement softening laws for the interface under mixed-mode loading condition

is assumed to support only the normal stress and shear stresses on the interface plane *x–y*, as illustrated in Fig. 7.6a. In pure debonding mode (opening Mode I, shearing Mode II or tearing Mode III), the onset of damage occurs when the tensile normal $\langle\sigma_{33}\rangle$ or shear traction (τ_{13}, or τ_{23}) attained their respective tensile strengths, *T* or shear strength, *S* of the interface, respectively, as illustrated in Fig. 7.5b, c. In a mixed-mode loading condition, the onset of damage due to interface debonding is assumed to occur when the following condition is satisfied [10, 11]:

$$\sqrt{\left(\frac{\langle\sigma_{33}\rangle}{T}\right)^2 + \left(\frac{\tau_{13}}{S}\right)^2 + \left(\frac{\tau_{23}}{S}\right)^2} = 1 \tag{7.6}$$

The term in Macaulay's bracket, $\langle\sigma_{33}\rangle$ is considered only when the normal stress is tensile. Compressive normal stress acting across the interface is assumed to maintain the interface intact, thus will not contribute to damage.

Based on the notations shown in Fig. 7.6a, the relative displacement for the mixed-mode loading can be written as:

$$\delta_m = \sqrt{\langle\delta_3\rangle^2 + \delta_{shear}^2} \tag{7.7}$$

The corresponding mixed-mode displacement at the onset of interface damage is expressed as:

$$\delta_m^o = \delta_1^o\delta_3^o\sqrt{\frac{1+\beta^2}{\left(\delta_1^o\right)^2 + \left(\beta\delta_3^o\right)^2}} \tag{7.8}$$

where the mode mixity term, β is defined as:

$$\beta = \frac{\delta_{shear}}{\delta_3}, \; \delta_3 > 0 \tag{7.9}$$

Following a damage initiation event, subsequent damage development is predicted based on fracture energy considerations [5, 13, 23, 24, 26]. An example of damage propagation criterion under mixed-model loading, expressed in terms of the total energy release rates, G_T and single mode critical energy release rates of the interface, G_I, G_{II} and G_{III} such that $G_T = G_I + G_{II} + G_{III}$ is given by [3]:

$$G_T = G_{IC} + (G_{IIC} - G_{IC})\left(\frac{G_{II} + G_{III}}{G_T}\right)^{\eta} \tag{7.10}$$

The exponent η accounts for the synergistic effect of mixed-mode fracture. The critical energy release rate for the isolated loading mode is represented by the area under the bilinear traction-displacement curve for the respective loading mode as (see also Fig. 7.5b, c):

$$G_{IC} = \frac{T\delta_n^f}{2} \tag{7.11a}$$

$$G_{IIC} = G_{IIIC} = \frac{S\delta_s^f}{2} \tag{7.11b}$$

These toughness properties of the interface should be experimentally determined. The energy release rate of the interface could be calculated by integrating the area under the traction-displacement curve for the respective loading mode up to the current loading condition.

The displacement of fracture under mixed-mode loading is expressed as:

$$\delta_m^f = \frac{2}{K_p\delta_m^o}\left[G_{IC} + (G_{IIC} - G_{IC})\left(\frac{\beta^2}{1+\beta^2}\right)^{\eta}\right] \tag{7.12}$$

The damage evolution for mixed-model loading conditions of the interface can then be expressed as:

$$D = \frac{\delta_m^f\left(\delta_m - \delta_m^o\right)}{\delta_m\left(\delta_m^f - \delta_m^o\right)} \tag{7.13}$$

The traction-relative displacement softening laws under mixed-mode loading condition are schematically illustrated in Fig. 7.6b.

The formulation of CZM for the interface failure process results in smoothing of the stress singularities associated with the crack tip. It also eliminates the need to separate crack initiation from crack propagation analysis.

Table 7.1 Parameters for cohesive zone model (CZM) of solder/IMC interfaces

Parameter	Description
k_n	Penalty stiffness (GPa)
k_s	Shear penalty stiffness (GPa)
T	Tensile strength (MPa)
S	Shear strength (MPa)
G_{IC}	Critical energy release rate for normal mode (N/mm)
G_{IIC}, G_{IIIC}	Critical energy release rate for shear mode (N/mm)
η	Mixed-mode fracture parameter

7.3.3 Determination of Model Parameter Values

Cohesive zone model parameters associated with the damage initiation criterion, Eq. (7.6), and damage evolution law, Eq. (7.10), for the interface are described in Table 7.1 [21, 34]. The relative displacement variable is normalized by the characteristic size of element at the interface, thus the penalty stiffness is expressed in similar fashion as local material stiffness. For simulation purposes, an interface of *zero thickness* is assumed to separate the bulk solder from the adjacent intermetallic (IMC) layer in leaded and lead-free solder joints. Tensile and shear strength properties of the solder/IMC interface are established through experiments on samples with reflowed solder-on-substrate joints. In view of the minute-size reflowed solder samples, direct tension and shear test setup for determination of solder/pad interface strengths is impractical. In this respect, a cold ball pull test and a solder ball shear push test on reflowed solder specimens are performed to generate tensile and shear strength data of the interface, respectively. Requirements for sample preparation and testing procedures are described in details in relevant testing standards [17, 18].

A schematic of the setup for the solder ball shear push test is illustrated in Fig. 7.7a. The shear tool is initially positioned with prescribed clearance between the tip of the tool and the surface of the substrate. However, the spherical shape of the bulk solder joint results in a larger shear tool-solder joint contact clearance, as illustrated in the figure. Movement of the shear tool imparts the force to the solder/IMC interface region causing fracture of the joint. The speed of the shear tool dictates the load transfer characteristic to the solder joint and solder/IMC interface. A slow speed will deform the solder, resulting in failure of the more compliant solder ball. Fast tool speed is desirable to facilitate rapid load transfer to the interface and initiate interface fracture. Properties of solder/IMC interface are expected to be strain rate-dependent because the properties of the adjacent solder are rate-dependent. The shear tool-solder contact clearance induces a significant bending effect at the solder/IMC interface during the test. Consequently, interface fracture initiates at the edge of the interface under combined shear and bending stresses. In this respect, hybrid experimental-computational approach can be employed to determine the shear strength of the solder/IMC interface. It has also been demonstrated that the average shear strength defined by the peak shear force over the measured sheared area corresponds to the minimum shear strength of the interface [35].

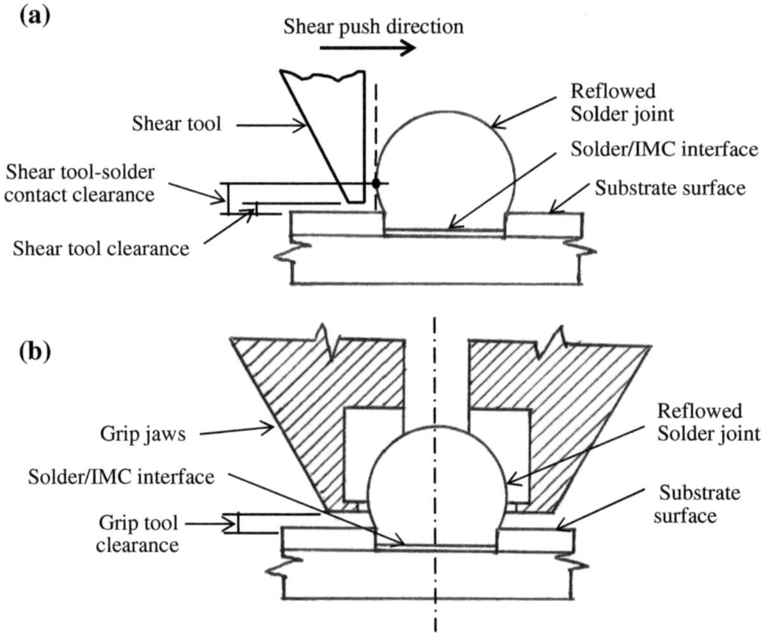

Fig. 7.7 **a** Set-up for solder ball shear push test and, **b** solder ball pull test

A solder ball pull test, commonly known as a cold ball pull test when performed at room temperature, is set up as illustrated in Fig. 7.7b. A grip clamps the solder ball with suitable pressure and pulls the joint upwards in the direction normal to the solder/IMC interface. It is postulated that the gripping of the solder does not induce any stress on the solder/IMC interface being tested. A fast pulling speed is required to ensure load transfer to the interface instead of deforming the bulk solder joint. The nominal tensile strength of the interface is determined by the maximum load at fracture over the fractured area at the interface. The fractured surface is examined to confirm that the crack traversed across the solder/IMC interface.

Both the solder ball shear push test and the pull test are performed with controlled displacement of the tool. The resisting tool load and displacement data pairs are recorded throughout the test until fracturing of the solder joint is detected. The resulting force-displacement curve is schematically shown in Fig. 7.8, comparing ductile failure and brittle fracture mode of the solder joint. In view of the brittle nature of solder/IMC interface, a brittle fracture mode is desired. The brittle fracture of the interface is characterized by an increasing force-displacement curve to the strength level of the interface followed by the sudden load drop due to the catastrophic fracture across the interface plane. Fractographic examination of the fractured interface should reveal the brittle fracture mechanism. The ductile failure mode of the solder joint is characterized by the attainment of the peak tool force, followed by a gradual decrease of the force over large tool displacement, as illustrated in Fig. 7.8. Ductile failure often results in a crack traversing in the bulk solder but close to the solder/IMC interface. Thus, the

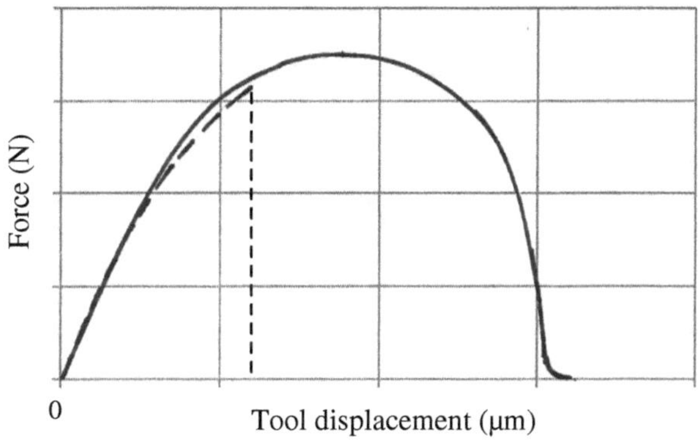

Fig. 7.8 Comparison of typical load-displacement curve for solder ball shear push test or ball pull test illustrating ductile failure mode (*solid line*) and brittle fracture mode (*dashed line*) of the solder joint

desirable interface fracture is not achieved. Mixed fracture conditions have also been observed in some solder/substrate systems. The resulting type of fracture depends on several factors, including the types of solder/substrate system, surface finishes, aging conditions, tool displacement rate and test temperature.

The measured force-tool displacement data for each test represents the deformation response of the reflowed solder system consisting of the bulk solder joint, solder/IMC interface and shear tool or grips. However, the desired information is the strength property of the solder/IMC interface. In this respect, the resulting fracture surface is analyzed to determine if the crack propagated across the solder/IMC interface plane or traversed across the bulk solder. For an interface fracture, a hybrid experimental-computational approach is then applied to the measured load-displacement curve to extract the intrinsic tensile and shear strength of the interface. The approach was described in Sect. 2.1 for the determination of interface shear strength using measured test data from solder ball shear push tests of reflowed Sn-40Pb solder joints. Similar procedures can be repeated for the solder ball pull test to establish the tensile strength of the interface. In addition, the cohesive zone model parameter, namely penalty stiffness and relative displacement at fracture can also be simultaneously extracted from the FE simulation results of the test.

A typical set of CZM parameter values for SAC405 and Sn-3.8Ag-0.7Cu solder joints are listed in Table 7.2.

In reliability assessment of solder joints under a board-level drop test, the BGA solder joints are likely to experience high strain rate loading. In this respect, strain rate dependency of solder/IMC interface properties should be quantified and incorporated into the cohesive zone model for FE simulation of solder joint deformation and the failure process. This is particularly critical in view of the observed decrease in interface strength with an increase in strain rate [7].

7.3.4 Extension of Cohesive Model for Cyclic Loading

In the relatively brittle solder/IMC interface, the cyclic interface damage process can be described based on the cohesive behavior of the interface [12, 43]. The bilinear traction-displacement law described earlier for monotonic loading can be

Table 7.2 CZM parameters values for solder/IMC interface in some Pb-free solder joints

Parameter	SAC405 [19]	SAC405 [39]	Sn-3.8Ag-0.7Cu [36]
k_n (GPa)	400	41	15
k_s (GPa)	400	15	15
T (MPa)	450	56.77	70
S (MPa)	450	43.66	70
G_{IC} (N/mm)	3	0.04	–
G_{IIC}, G_{IIIC} (N/mm)	3	0.065	–
η	1.45	1	–

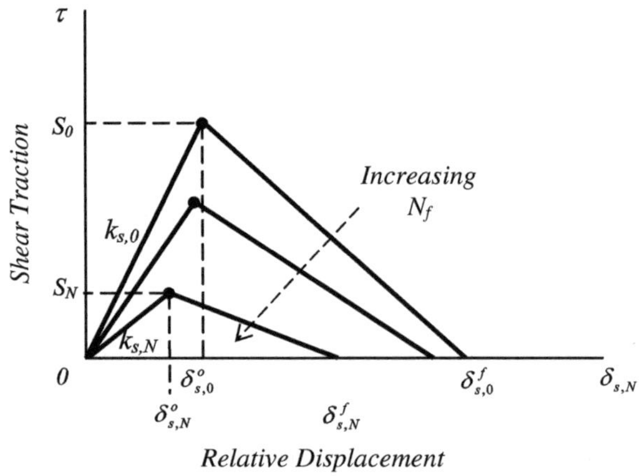

Fig. 7.9 Fatigue traction-displacement curves illustrating cyclic degradation of residual interface properties for shear loading mode

extended to fatigue loading through consideration of cyclic degradation of interface properties. An example of fatigue traction-displacement curves illustrating cyclic degradation of the five cohesive zone model parameters (see Table 7.1) for single mode loading is schematically shown in Fig. 7.9. This figure also represents the decreasing residual strength of the interface after enduring cyclic stressing, N_f as depicted by the lower residual bilinear traction-displacement curves.

In view of the complexity of the determination of cyclic cohesive zone model parameter (CCZM) values, it is assumed as a first approximation that only selected cohesive interface properties, namely interface strength (S, T), penalty stiffness (k_s, k_n) and critical strain energy release rate (G_{IC}, G_{IIC}) are influenced by fatigue [40]. It is further assumed that each relative displacement corresponding to the onset of fatigue damage and final separation of the material point is independent of the number of endured load cycles. The proposed fatigue traction-separation law for the cohesive interface is illustrated in Fig. 7.10 for shear component. Similar cohesive behavior is described for the normal stress component to the interface plane.

The progressive cyclic degradation of the interface is quantified in terms of interface damage parameters, D_{cyc} using an evolution law of the assumed form:

$$D_{cyc} = 1 - (N)^b \tag{7.14}$$

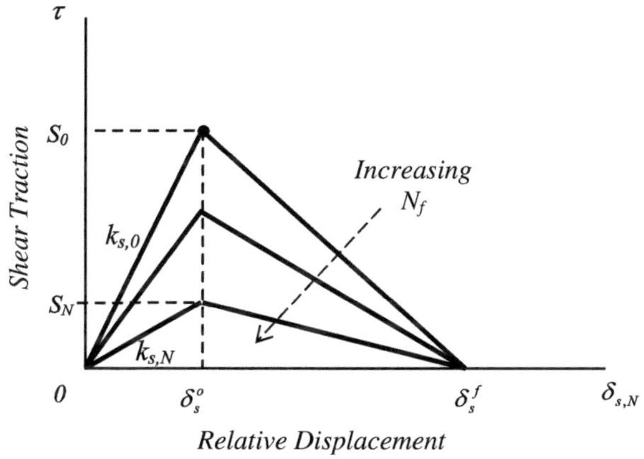

Fig. 7.10 Simplified fatigue traction-separation law for cyclic shear loading of solder/IMC interface

Table 7.3 Parameters for cyclic cohesive zone model (CCZM) of solder/IMC interface

Initial parameter, H_0	Fatigue parameter, H_{cyc}	Description
$k_{n,0}$	$k_{n,N}$	Residual penalty stiffness (GPa)
$k_{s,0}$	$k_{s,N}$	Residual shear penalty stiffness (GPa)
T_0	T_N	Residual tensile strength (MPa)
S_0	S_N	Residual shear strength (MPa)
$G_{IC,0}$	$G_{IC,N}$	Fatigue critical energy release rate for normal mode (N/mm)
$G_{IIC,0}$, $G_{IIIC,0}$	$G_{IIC,N}$, $G_{IIIC,N}$	Fatigue critical energy release rate for shear mode (N/mm)
η		Mixed-mode parameter
b		Fatigue damage exponent

The initial parameters, H_0 are identical to that shown in Tables 7.1 and 7.2

where the exponent, b is an experimentally determined fatigue damage parameter. Thus, cyclic degradation of each interface property, H_{cyc} can be estimated as:

$$H_{cyc} = H_0\left(1 - D_{cyc}\right) = H_0(N)^b \tag{7.15}$$

Interface properties represented by H_{cyc} for CCZM, as illustrated in Fig. 7.10, are summarized in Table 7.3. The coefficient, H_0 represents the initial (monotonic) value of the interface property of penalty stiffness, $\left(k_{S,0}, k_{n,0}\right)$ interface strengths, (S_0, T_0) and critical energy release rates, $\left(G_{IC,0}, G_{IIC,0}\right)$. These properties can be extracted from measured data on a series of cold ball pull tests and ball shear push tests on reflowed solder-on-substrate samples.

In the absence or limited availability of fatigue life data for solder/IMC interface in open literature, the assumed form of fatigue degradation of interface properties (Eq. 7.15) is illustrated based on the strength-life (S–N) curve of SAC405

Fig. 7.11 Fatigue behavior of SAC405 solder joints (data are extracted from Khatibi et al. [21])

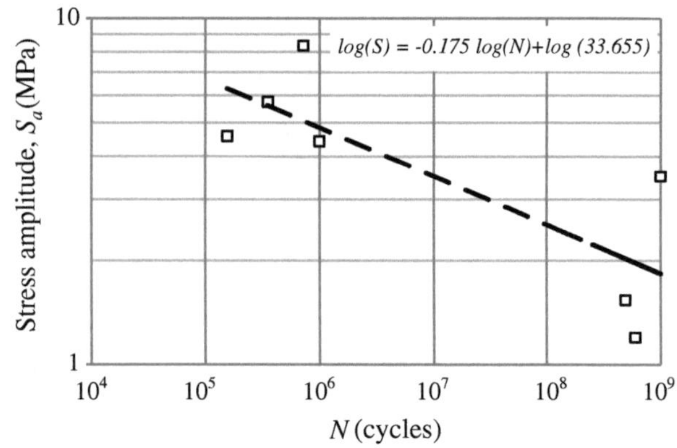

Fig. 7.12 **a** Characteristic cyclic damage evolution and, **b** cyclic property degradation for SAC405/Cu_6Sn_5 interface system with fatigue damage exponent, $b = -0.175$

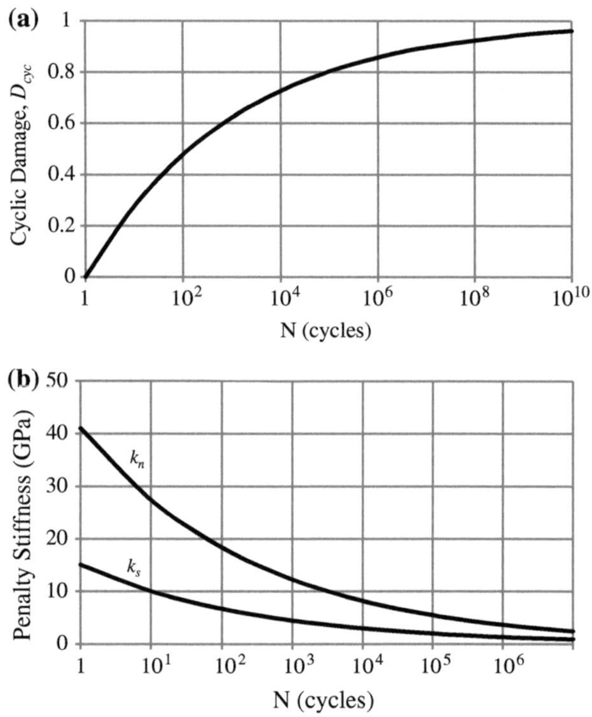

solder joints [21]. The best-fit of such limited data results in the fatigue exponent value of $b = -0.175$, as shown in Fig. 7.11. Similar characteristic fatigue degradation of other interface properties is assumed and employed in case studies, as discussed in Chaps. 8 and 9. The resulting characteristic evolution of cyclic damage (Eq. 7.14) is shown in Fig. 7.12a, while an example of cyclic degradation of interface properties (penalty stiffness) is shown in Fig. 7.12b for SAC405/Cu_6Sn_5 interface. Since the fatigue damage exponent governs the evolution of damage, thus dictates the fatigue life of the solder joint through the fracture of solder/IMC interface, a better set of experimental data on the fatigue of solder/IMC interface is required. A better form of the cyclic damage evolution law would be possible upon availability of measured fatigue-related variables of the interface.

The updated traction at a material point of the solder/IMC interface can then be expressed as:

$$\begin{aligned} \sigma &= (N)^b k_{n,0} \delta_n \\ \tau &= (N)^b k_{s,0} \delta_{shear} \end{aligned} \tag{7.16}$$

The stress-based criterion for the onset of damage and energy-based criterion for damage progression defined by Eqs. (7.6) and (7.10) respectively are employed with cycle-degraded interface properties. Applied cyclic loading continuously degrades these interface property values while simultaneously dictating interface damage evolution. Material damage accumulates throughout the cyclic loading to a specified terminal level at the fracture of the interface.

Other form of cohesive zone models accounting for stress reversals, each with different assumptions on unloading-reloading paths [27, 29, 41, 42] and cyclic damage evolution laws [14, 32, 43] have been introduced and examined.

7.4 Continuum Damage Model

A crack initiated in the vicinity of solder/IMC interface of a solder joint could also propagate under fatigue loading into the bulk solder. In this respect, a continuum damage-based model is required to quantify the progressive damage and fatigue failure process of the bulk solder joint. A similar concept of continuum damage of a material point under monotonic loading (see Fig. 7.4) is used to describe fatigue damage initiation and propagation in the bulk solder joint.

7.4.1 Cyclic Continuum Damage Model

The onset of cyclic damage in the bulk solder is characterized by the accumulated inelastic hysteresis energy per stabilized cycle, ΔW, calculated as the volume averaged quantity over the critical region. Fatigue cycles to damage initiation, N_o is represented by [9]:

$$N_0 = C_1 (\Delta W)^{C_2} \tag{7.17}$$

where C_1 and C_2 are material constants. Following damage initiation, fatigue damage evolution of the bulk solder is quantified based on the characteristic crack propagation rate. The continuum damage evolution rule is assumed to be in the following form:

$$\frac{dD}{dN} = \frac{C_3 (\Delta W)^{C_4}}{L} \tag{7.18}$$

Table 7.4 Material constants for cyclic continuum damage model (CCDM) for SAC405 solder joint (based on data extracted from Tunga [37])

Parameter	$C1$	$C2$	$C3$	$C4$
Value	22.55	−1.55	4.94×10^{-4}	0.98

where the *LHS* damage evolution term represents the damage propagation rate in the bulk solder joint. The parameters C_3 and C_4 are material constants, and L is the characteristic length of the damaged element. It is noted that the fatigue parameter representing the accumulated inelastic hysteresis energy per stabilized cycle, ΔW is identical to that used in the classical fatigue model, $W_{in,acc}$ as given by Eqs. (4.32) and (4.33). Typical material constants extracted from published data for SAC405 solder are listed in Table 7.4 [37].

The current stress tensor, σ is updated such that:

$$\sigma = (1 - D)\bar{\sigma} \tag{7.19}$$

where $\bar{\sigma}$ is the effective stress tensor that would exist in the material in the absence of damage, computed in the current load increment. It is noted that the solder would have completely lost its load carrying capacity when $D = 1$ and separation of the material point occurs. Collection of these separated material points is interpreted in this simulation as representing the traction free crack surfaces.

References

1. Anderson, T.L.: Fracture Mechanics: Fundamentals and Applications. CRC Press, Boca Raton (2005)
2. Barenblatt, G.I.: The mathematical theory of equilibrium cracks in brittle fracture. Adv. Appl. Mech. **7**, 55–129 (1962)
3. Benzeggah, M.L., Kenane, M.: Measurement of mixed-mode delamination fracture toughness of unidirectional glass/epoxy composites with mixed-mode bending apparatus. Compos. Sci. Technol. **49**, 439–449 (1996)
4. Broek, D.: Elementary Engineering Fracture Mechanics. Noordhoff International Publishing, Leyden (1974)
5. Camanho, P.P., Matthews, F.L.: Delamination onset prediction in mechanically fastened joints in composite laminates. J. Compos. Mater. **33**, 906–927 (1999)
6. Chaboche, J.L.: Anisotropic creep damage in the framework of continuum damage mechanics. Nucl. Eng. Technol. **79**, 309–319 (1984)
7. Darveaux, R., Corey, R.: Ductile-to-brittle transition strain rate. In: Proceedings Electronics Packaging Technology Conference (EPTC), pp. 283–289 (2006)
8. Darveaux, R., Reichman,C.: Ductile to brittle transition rate. In: 8th Electronics Packaging Technology Conference Proceedings, pp. 283–289 (2006)
9. Darveaux, R.: Effect of simulation methodology on solder joint crack growth correlation and fatigue life prediction. J. Electron. Packag. **124**, 147–154 (2002)
10. Dávila, C.G., Camanho, P.P., Moura, M.F.: Mixed-mode decohesion elements for analyses of progressive delamination. In: Proceedings of 42nd AIAA/ASME/ASCE/AHS/ASC Structures, Structural Dynamics and Materials Conference, pp. 16–19, Seattle, WA, USA (2001)

11. Dávila, C.G., Camanho, P.P.: Analysis of the effects of residual strains and defects on skin/stiffener debonding using decohesion elements. In: Proceeding of 44th AIAA/ASME/ASCE/AHS Structures, Structural Dynamics, and Materials Conference. Norfolk, VA (2003)
12. Dávila, C.G., Camanho, P.P.: Decohesion elements using two and three-parameter mixed-mode criteria. In: Proceedings of American Helicopter Society Conference, (2001)
13. Davila, C.G., Johnson, E.R.: Analysis of delamination initiation in postbuckled dropped-ply laminates. AIAA J. **31**(4), 721–727 (1993)
14. De-Andrés, A., et al.: Elastoplastic finite element analysis of three-dimensional fatigue crack growth in aluminum shafts subjected to axial loading. Int. J. Solids Struct. **36**(15), 2231–2258 (1999)
15. Dugdale, D.S.: Yielding of steel sheets containing slits. J. Mech. Phys. Solids **8**(2), 100–104 (1960)
16. Gurson, A.L.: Continuum theory of ductile rupture by void nucleation and growth: Part I Yield criteria and flow rules for porous ductile media. J. Eng. Mater. Technol. Trans. ASME **99**, 2–15 (1977)
17. JEDEC: JESD22-B115A, Solder Ball Pull (2010)
18. JEDEC: JESD22-B117A, Solder Ball Shear, Arlington, VA (2006)
19. Jing, J., et al.: Simulation of dynamic fracture along solder-pad interfaces using a cohesive zone model. In: ASME Conference Proceedings (48678), pp. 171–176 (2008)
20. Kachanov, L.M.: Introduction to Continuum Damage Mechanics. Martinus Nijhoff, Dordrecht (1986)
21. Khatibi, G., et al.: A novel accelerated test technique for assessment of mechanical reliability of solder interconnects. Microelectron. Reliab. **49**(9–11), 1283–1287 (2009)
22. Krueger, R., Cvitkovich, M.K., O'Brien, T.K., Minguet, P.J.: Testing and analysis of composite skin/stringer debonding under multi-axial loading. Compos. Mater. **34**, 1263–1300 (2000)
23. Krueger, R., Minguet, P.J., O'Brien, T.K.: A method for calculating strain energy release rates in preliminary design of composite skin/stringer debonding under multi-axial loading. NASA TM-1999-209365 (1999)
24. Krueger, R., París, I.L., O'Brien, T.K., Minguet, P.J.: Fatigue life methodology for bonded composite skin/stringer configurations. NASA/TM-2001-210842. VA (April 2001)
25. Lemaitre, J.: A Course on Damage Mechanics. Springer, Berlin (1992). 2nd Edition (1996)
26. Li, J., Sen, J.K.: Analysis of frame-to-skin joint pull-off tests and prediction of the delamination failure. In: Proceedings of 42nd AIAA/ASME/ASCE/AHS/ASC Structures, Structural Dynamics and Materials Conference, Seattle, WA (2000)
27. Li, Y.N., et al.: Theory of cohesive crack model with interactive cracks. Int. J. Solids Struct. **35**(11), 981–994 (1998)
28. Murakami, S.: Continuum Damage Mechanics: Solid Mechanics and its Applications. Springer, Berlin (2012)
29. Ortiz, M.: Microcrack coalescence and macroscopic crack growth initiation in brittle solids. Int. J. Solids Struct. **24**(28–29), 231–250 (1988)
30. Rice, J.R.: Mathematical analysis in the mechanics of fracture. Fracture **2**, 191–311 (1968)
31. Robotnov, Y.N.: Creep Rupture, in Proceedings of Applied Mechanics Conference, H.M. Vincenti (eds). Stanford University, Springer, Berlin. 342-349 (1968)
32. Roe, K.L., Siegmund, T.: An Irreversible cohesive zone model for interface fatigue crack growth simulation. Eng. Fract. Mech. **70**(2), 209–232 (2003)
33. Rousselier, G.: Ductile fracture models and their potential in local approach of fracture. Nucl. Eng. Des. **105**, 97–111 (1987)
34. Song, F.: Experimental Investigation on Testing Conditions of Solder Ball Shear and Pull Tests and the Correlation with Board Level Mechanical Drop Test. University of Science and Technology, Hong Kong (2007)
35. Tamin, M.N., Nor, F.M., Loh, W.K.: Hybrid experimental-computational approach for solder/IMC interface shear strength determination in solder joints. IEEE Trans. Compon. Packag. Technol. **33**(3), 614–620 (2010)
36. Towashiraporn, P., Xie, C.: Cohesive modeling of solder interconnect failure in board level drop test. In: The 10th Intersociety Conference on Thermal and Thermomechanical Phenomena in Electronics Systems (ITHERM '06) (2006)

37. Tunga, K.R.: Study of Sn-Ag-Cu alloy reliability through material microstructure evolution and laser moire interferometry. Ph.D. Thesis, School of Mechanical Engineering, Georgia Institute of Technology, USA (2008)
38. Turon, A., Costa, J., Camanho, P.P., Dávila, C.G.: Simulation of delamination in composites under high-cycle fatigue. Composites **38**, 2270–2282 (2007)
39. Yamin, A.F.M., Shaffiar, N.M., Loh, W.K., Tamin, M.N.: Extended cohesive zone model for simulation of solder/IMC interface cyclic damage process in Pb-free solder interconnects. In: Proceedings of 35th International, Electronic Manufacturing Technology Conference, IEMT (2012)
40. Yamin, A.F.M: Dynamic fracture process of solder/intermetallic interface in lead-free solder interconnects using cohesive zone model. Master Thesis, Universiti Teknologi Malaysia, Malaysia (2012)
41. Yang, B., et al.: A cohesive zone model for fatigue crack growth in quasibrittle materials. Int. J. Solids Struct. **38**(22–23), 3927–3944 (2001)
42. Yang, B., Ravi-Chandar, K.: A single-domain Dual-boundary-element formulation incorporating a cohesive zone model for elastostatic cracks. Int. J. Fract. **93**(1–4), 115–144 (1998)
43. Yang, Q.D., et al.: A cohesive zone model for low cycle fatigue life prediction of solder joints. Microelectron. Eng. **75**(1), 85–95 (2004)

Chapter 8
Application III: Board-Level Drop Test

8.1 Introduction

Consumer electronic products, particularly mobile devices such as mobile phones, thermal scanners, tablets and laptop computers, contain BGA packages and assemblies. These devices are susceptible to impact loading from unintentional dropping of the device during operation and handling. Such loading transfers the induced mechanical shock to solder joints in the electronic assembly that could initiate damage, leading to premature failure of the solder joints. Consequently, mechanical shock loading is considered in reliability assessment of solder joints. Impact loading is introduced through a board-level drop test of a PCB with a surface mounted electronic assembly containing BGA solder joints. The accelerated and controlled drop test is intended to reproduce an identical deformation response and failure mechanisms of critical solder joints in the test assembly, as would likely to occur in the actual product drop event. Thorough understanding of the fracture process of the solder joint, both the mechanics and mechanisms of bulk solder failure and solder/IMC interface fracture, contributes directly in quantifying the reliability of newly designed electronic packages and assemblies with respect to impact loading.

This chapter describes the systematic procedures in assessing reliability of solder joints in a BGA assembly subjected to impact loading conditions from board-level drop testing of the component. A standard board-level drop test of a PCB with BGA assembly is described. The vibration response of the board and characteristic deformation of critical solder joints in the assembly are quantified through FE simulation of the test. Procedures for setting up a mathematical model of the problem, developing the geometrical model for the analysis and prescribing a damage model for the solder/IMC interface are deliberated. Results are interpreted and discussed with respect to the mechanics of deformation and brittle fracture of the solder/IMC interface.

M. N. Tamin and N. M. Shaffiar, *Solder Joint Reliability Assessment*, Advanced Structured Materials 37, DOI: 10.1007/978-3-319-00092-3_8,

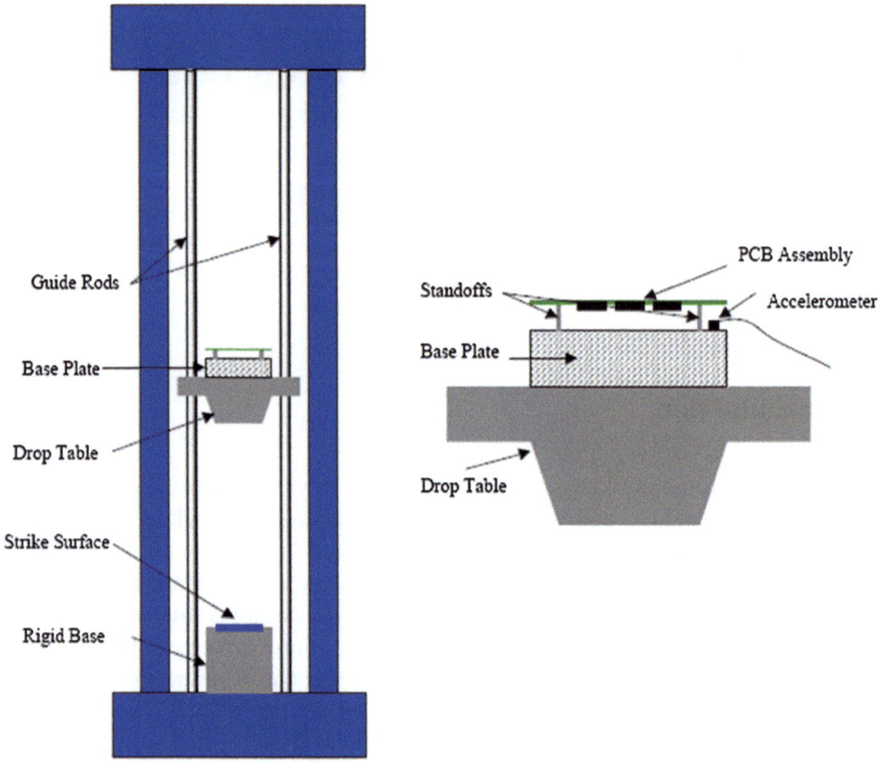

Fig. 8.1 A typical setup of a board-level drop test [1]

8.2 Simulation of Board-Level Drop Test

Requirements and procedures for conducting a board-level drop test on electronic packages and assemblies are described in JEDEC Test Standards [1, 2]. The test is designed to evaluate and compare the drop performance of newly-design products under an impact environment where excessive flexing of the circuit board could cause product failure. A typical setup of the test is illustrated in Fig. 8.1. The PCB assembly (with the attached electronic package face facing downwards) is mounted at each corner to a base plate that is secured to the drop table. The table is raised to a prescribed height and then released to freely impact the strike surface of a rigid base repetitively. An accelerometer attached to the base plate measures the acceleration (G-level, pulse duration and shape) at impact while a dynamic load cell records the load. The impact induces stress waves that travel through the mounting points of the PCB assembly to the BGA solder joints located in the center of the board. A failure analysis is performed on the damaged electronic assembly by combinations of visual inspection, cross-sectioning, dye-and-pry, chemical etching and scanning microscopy. The analysis aims to quantify the extent of impact failure and identify the associated fracture mechanisms of the BGA solder joints. Since the strength of the relatively brittle solder/IMC interface typically decreases with increasing strain rates, it is expected that fracture of the interface will dominate the failure process. A dominant solder/IMC interface fracture was observed for aged SAC405/Ni-coated copper substrate system for a strain rate in excess of 0.03 s^{-1} [3].

Finite element simulation of the board-level drop test is developed to accurately reproduce the deformation response of the PCB assembly during the test. Vibration analysis of the PCB assembly identifies the natural frequencies and possible mode shapes of the assembly. Subsequent FE simulation provides vital information on the internal states of displacement, strain and stress in the electronic assembly, particularly in the solder joints. In addition, a characteristic evolution of inelastic strain and damage in critical solder joints of the array is calculated. Such information relates to the reliability assessment of the solder joints under impact loading conditions.

8.2.1 Mathematical Model

The test board used in the simulation measures 132 mm × 77 mm with a thickness of 1 mm [1]. An electronic assembly with a BGA package contains a 5.6 mm × 5.6 mm × 0.33 mm silicon die and a 10 mm × 10 mm × 0.26 mm substrate mounted in the center of the board. An array of 10 × 10 solder joints are modeled as solder mask-defined (SMD) at the device side and non-solder mask-defined (NSMD) at the board side of the assembly. The Sn-4Ag-0.5Cu (SAC405) solder joints, each with a diameter of 0.48 mm and stand-off height of 0.3 mm, are arranged in an area array with a pitch distance of 0.8 mm. A thin layer (2 μm) of Cu_6Sn_5 intermetallic is assumed to have developed at the solder/pad interface.

The full model geometry of the PCB assembly is employed in modal analysis. The vibration of the assembly that is fixed at the four mounting points is simulated to establish natural frequencies and the corresponding model shapes. Based on the predicted first five mode shapes and natural frequencies of the board, the first mode with the most severe bending of the electronic assembly is considered (refer to discussion of results on modal analysis below). The resulting deformation is symmetry about both center lines of the PCB assembly; only one-quarter of the model geometry is sufficient for consideration of subsequent stress analysis. The quarter PCB assembly model is illustrated in Fig. 8.2. Symmetry boundary conditions are prescribed to the *xz*- and *yz*-plane as follows:

$$\text{Symmetry about } xz\text{-plane: Displacement, } U_y = 0 \text{ and rotation, } U_{Rx} = U_{Rz} = 0$$
$$\text{Symmetry about } yz\text{-plane: Displacement, } U_x = 0 \text{ and rotation, } U_{Ry} = U_{Rz} = 0$$

All surfaces initially in contact are assumed to be perfectly bonded throughout the simulation for the vibration response of the assembly. However, in subsequent

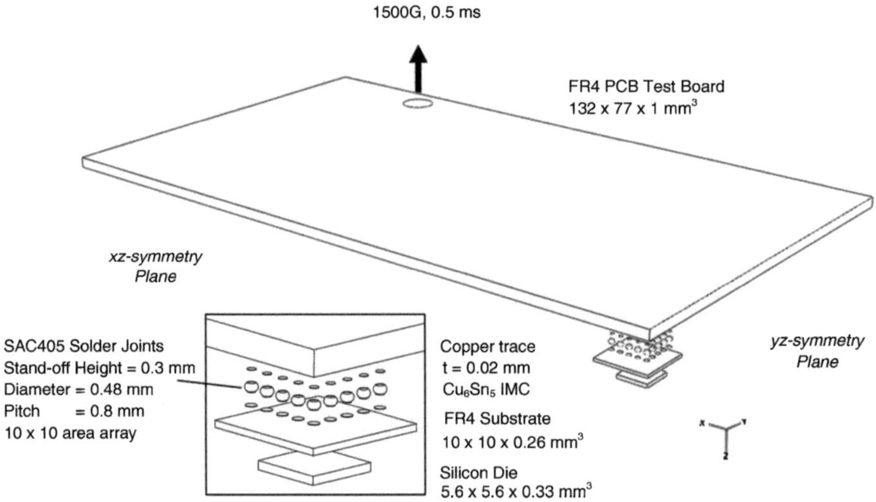

Fig. 8.2 Exploded view of the quarter PCB assembly model for board-level drop test simulation

refined stress analysis, the calculated damage may render separation of the critical solder/IMC interface.

The unified inelastic strain constitutive model (Anand model) [4] is employed to describe the strain rate-dependent response of SAC405 solder joints. The material model was described in Sect. 4.5 along with an optimized set of modified Anand model parameters [5]. The material model parameters for the solder as used in the simulation are reproduced for convenience in Table 8.1. In addition, orthotropic behavior of the FR-4 substrate and test board is also considered. Other materials are assumed to behave elastically throughout the simulations.

The damage behavior of the solder/IMC interface of the solder joints is represented by the cohesive zone model, with the bilinear traction-displacement softening law, as described in Sect. 7.3. However, it is assumed in this case study that the interface behavior is ideally brittle such that separation of simulated interface material points occurs instantaneously upon damage initiation. The cohesive zone model parameters for SAC405/Cu_6Sn_5 interface employed in this model are reproduced in Table 8.1c for convenience.

The physical board-level drop test can be simulated by releasing the drop-table from the actual drop height or from a close distance to the strike surface, but with a velocity equivalent to the actual free-fall velocity from the specified drop height. The drop-table will freely impact on the strike surface repetitively. However, an alternate Input-G method is employed in this simulation case [6]. In this simulation, it is assumed that each strike creates a half-sine impact acceleration pulse at the mounting points of the PCB assembly, approximated as:

$$G(t) = G_m \sin \frac{\pi t}{t_w} \tag{8.1}$$

Table 8.1 Materials properties and material model parameter values used in the simulation of the board-level drop test

Parameter	Description	Values
(a) *Unified inelastic strain model (Anand model) with refined model parameters*		
Q/R	Activation energy term	10,561 K
A	Pre-exponential factor	7,128 s^{-1}
ξ	Stress multiplier	10
m	Strain rate sensitivity of stress	0.8181
$\hat{s}$	Coefficient for deformation resistance saturation value	30.45 MPa
n	Strain rate sensitivity of the saturation value	1.091×10^{-3}
a	Strain rate sensitivity of the hardening coefficient	1.6

Refined parameter	a_0	a_1	a_2	a_3	a_4
s_0	25.996	−0.014772	7.386×10^{-6}	–	–
h_0	9.85×10^4	−880	2.2	-2.1×10^6	1.05×10^8

Material	SAC405	Cu layer	Cu_6Sn_5	Si-die
(b) *Properties for other materials used in the model* [1, 6, 11, 13]				
E (GPa)	41.05	128.53	85.60	129.57
ν	0.36	0.35	0.31	0.28

E (GPa)	*G* (GPa)	ν
FR-4 substrate (xy-plane is the isotropic plane)		
16.898 (*xy*)	7.632 (*xy*)	0.11 (*xy*)
7.436 (*z*)	3.325 (*xz, yz*)	0.39 (*xz, yz*)

Parameter	Description	Value
(c) *Cohesive zone model parameter values for SAC405/Cu_6Sn_5 interface*		
k_n	Penalty stiffness (GPa)	41
k_s	Shear penalty stiffness (GPa)	15
T	Tensile strength (MPa)	56.77
S	Shear strength (MPa)	43.66
G_{IC}	Critical energy release rate for normal mode (N/mm)	0.04
G_{IIC}, G_{IIIC}	Critical energy release rate for shear mode (N/mm)	0.065
η	Mixed-mode fracture parameter	1

$s_0 = 25.966 - 0.014772\,T + 7.386 \times 10^{-6}\,T^2$

$h_0 = 9.85 \times 10^4 - 880T + 2.2\,T^2 - 2.1 \times 10^6\,\dot{\varepsilon}_p + 1.05 \times 10^8\,\dot{\varepsilon}_p^2$

where G_m is the peak acceleration and t_w is the pulse duration. The peak acceleration can be calculated, assuming that the potential energy is fully converted into kinetic energy as:

$$G_m = \frac{\pi\sqrt{gH}}{t_w\sqrt{2}} \tag{8.2}$$

where $g = 9.81$ m/s^2 is the gravitational acceleration and H is the drop height.

Different equivalent drop heights represent different service conditions for the simulation. Table 8.2 lists various designated service conditions and the

Table 8.2 Board-level drop test parameters for different prescribed service conditions

Service condition	Equivalent drop height (inches)/(cm)	Velocity change (in/s)/(cm/s)	Acceleration peak (G)	Pulse duration (ms)
H	59/150	214/543	2,900	0.3
G	51/130	199/505	2,000	0.4
B	44/112	184/467	1,500	0.5
F	30/76.2	152/386	900	0.7
A	20/50.8	124/316	500	1.0
E	13/33.0	100/25.4	340	1.2
D	7/17.8	73.6/187	200	1.5
C	3/7.62	48.1/122	100	2.0

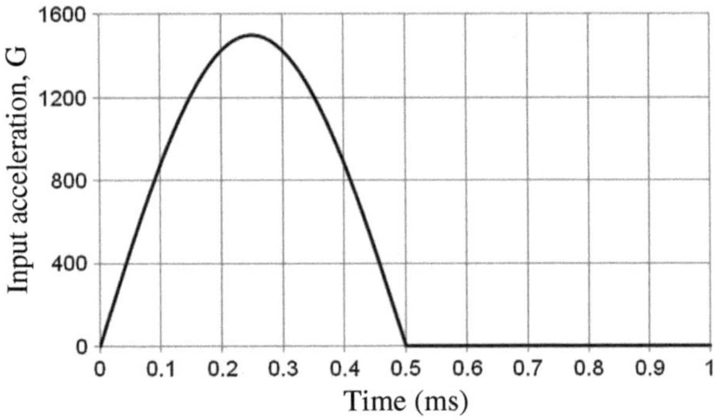

Fig. 8.3 Input pulse, expressed in multiples of gravitational acceleration for the equivalent drop height of $H = 1.12$ m

corresponding equivalent drop heights, and other loading parameters [1]. This case study simulates service condition B with an equivalent drop height of $H = 1.12$ m and pulse duration of $t_w = 0.5$ ms. The corresponding velocity and peak acceleration at impact is 4.67 ms^{-1} and 1,500g (expressed as 1,500G) or 14,715 ms^{-2} respectively. The input pulse loading for this service condition is illustrated in Fig. 8.3.

8.3 Finite Element Modeling

The quarter model of the PCB assembly is discretized into 8-node continuum elements with reduced integration points. The FE mesh of the equivalent Si-package, substrate, BGA solder joints and PCB is illustrated in Fig. 8.4a. The detailed mesh of the solder joints in the array with an element edge length of 0.05 mm is illustrated in Fig. 8.5b. A layer of cohesive elements is defined between the solder and Cu_6Sn_5 IMC layer on both device and board sides of the assembly.

A preliminary run of the PCB assembly model with this coarse FE mesh of the solder joints (edge length of 0.05 mm) has indicated that the corner solder joint

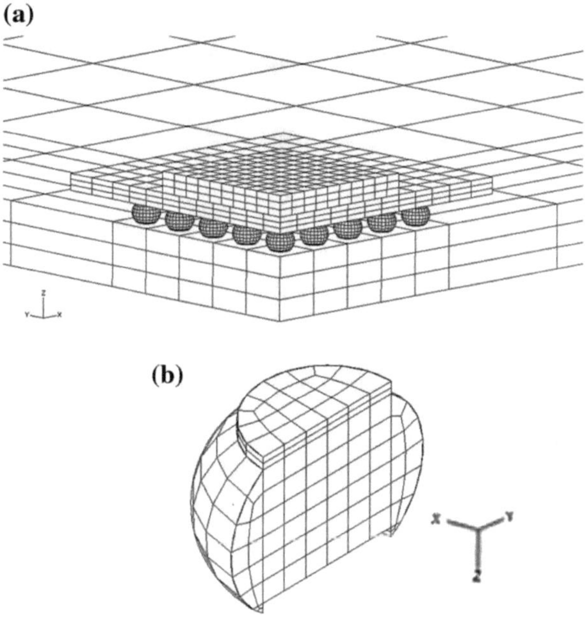

Fig. 8.4 **a** Finite element mesh of the quarter PCB assembly model with BGA solder joints. **b** Detailed coarse mesh of critical solder joints (cut out section)

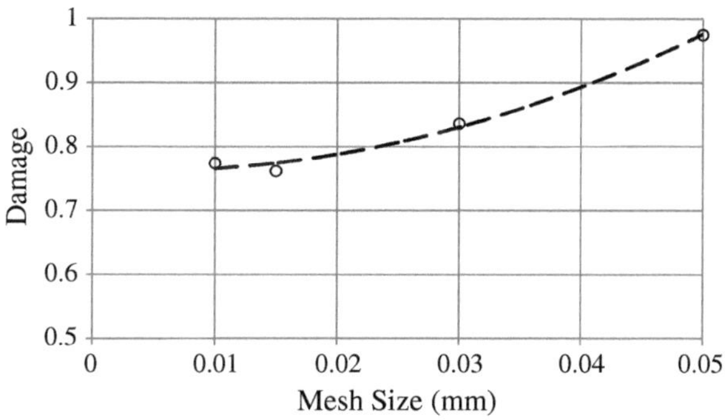

Fig. 8.5 Results of the mesh convergence study of elements in a solder/IMC interface

is the most critical with respect to damage accumulation. In the subsequent mesh convergence study, the size of elements in this solder joint is reduced successively for each simulation run while monitoring damage initiation variable at the selected location in the solder/IMC interface. The difference in the calculated magnitude of the damage variable between two successive models decreases with a decreasing size of elements, as shown in Fig. 8.5. This mesh convergence result suggests that the critical solder joint should be discretized into elements with an element edge length of 0.015 mm or smaller to ensure mesh-independent results. Elements in a solder/IMC interface with an edge length of 0.01 mm is then employed in subsequent simulations of the board-level drop test. In view of the potential propagation of the solder/IMC interface damage to neighboring solder joints, an additional

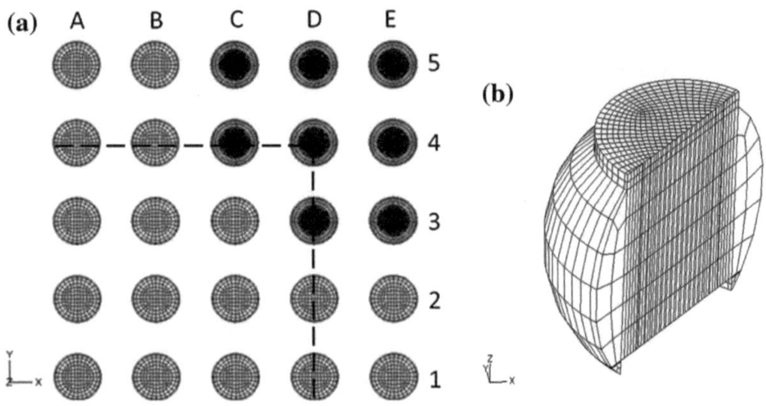

Fig. 8.6 **a** Expected location of critical solder joints in BGA solders (indicated by *dark color*). **b** Fine element mesh for the critical solder joint (cut out section)

seven solder joints in the immediate vicinity are also discretized into fine element mesh. The fine element mesh for selected solder joints in the BGA solders of the PCB assembly is illustrated in Fig. 8.6.

In this case study, each location of the solder joint in the array is identified using row (1–5) and column (A–E) identifiers. The dashed line overlays the edge of the stiff Si-package or die. Solder joint A1 is located closest to the center of the array (and PCB assembly) while D4 lies directly underneath the corner of the Si-die. Solder joints located along the diagonal line of the model (A1, B2, C3, D4 and E5) are of particular interest in this simulation.

8.4 Interface Damage Initiation and Dynamic Propagation

8.4.1 Modal Analysis of the PCB Assembly

The vibration characteristic of the PCB assembly is quantified in terms of natural frequencies and the corresponding mode shapes. The predicted first five natural frequencies and the corresponding mode shapes are illustrated in Fig. 8.7. The first mode (Mode 1) that is likely to occur represents bending of the board and is predicted to occur at 218.83 Hz. Previous experimental investigation shows the dominant frequency at the center of the board is around 200 Hz [7, 8]. This result leads to subsequent FE structural analysis to consider only the first mode shape with severe bending of the electronic assembly. It is noted that the associated deformation is symmetrical with respect to the two central axes of the geometry of the PCB test assembly. This justifies that one-quarter section of the PCB assembly is sufficient for consideration in subsequent FE simulation of the board-level drop test, as discussed above.

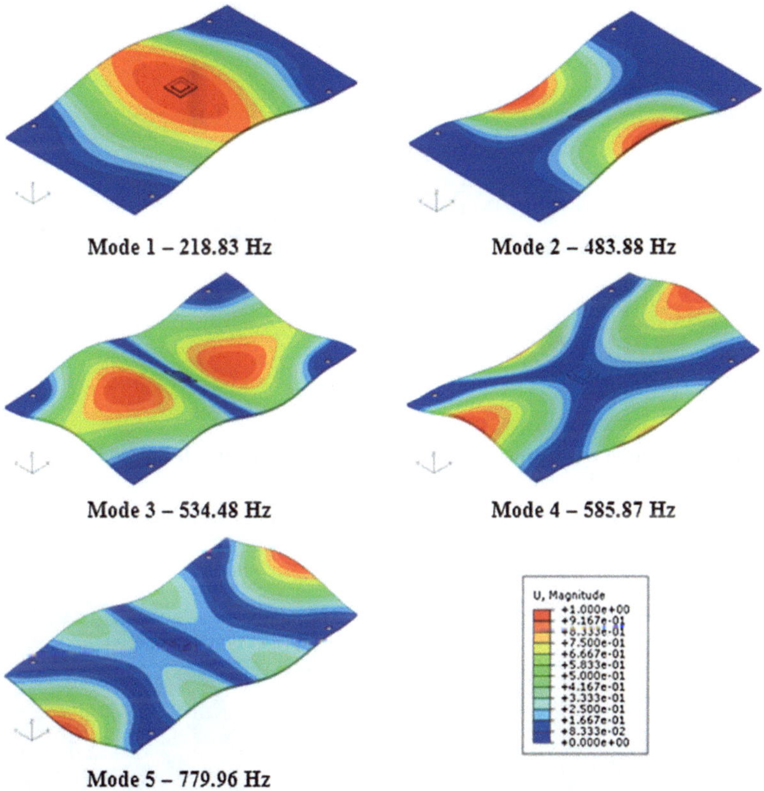

Fig. 8.7 First five natural frequencies and the corresponding mode shapes of the PCB assembly

8.4.2 Deflection of the Test Board

The predicted deformation of the PCB assembly at the end of the applied load pulse is illustrated in Fig. 8.8a. It is noted that the out-of-plane displacement contour (U_Z) is identical to that predicted by the full model of the PCB assembly in the vibration simulation (Refer to Fig. 8.7 for Mode 1, 218.83 Hz). This ensures correct representation of the boundary conditions for the quarter-model of the assembly. The constrained relative out-of-plane displacement at the center of the test board is 1.206 mm. Such a deformed contour of the board induces tensile bending stresses in the BGA assembly. It is found that the most critical solder joint with respect to inelastic strain in the solder and damage of the solder/IMC interface is the one located farthest away from the center of the package (E5), as illustrated in Fig. 8.8b. The solid line overlays the edge of the stiff Si-package. This solder joint experienced the largest in-plane (x–y) relative displacement during deformation and induced large shear stresses on the solder/IMC interfaces.

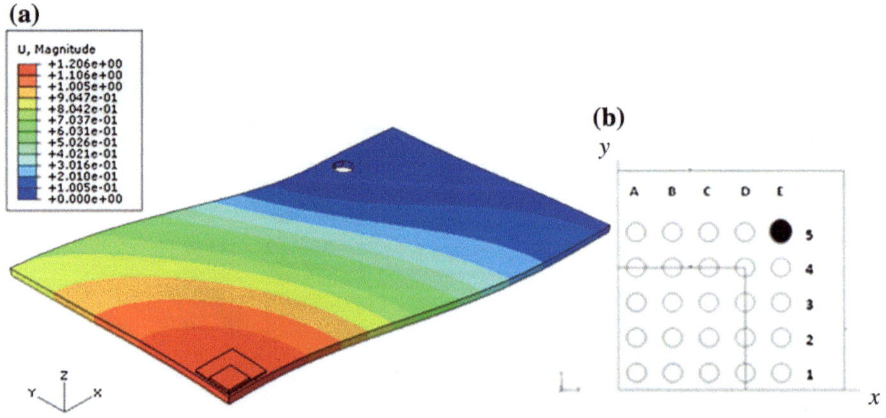

Fig. 8.8 **a** Deformation for the quarter-model of the PCB assembly, **b** location of the most critical solder joint in array (shown by *dark color*)

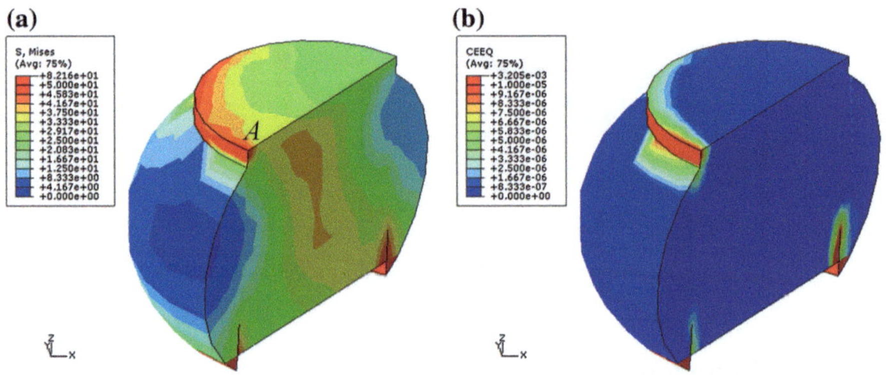

Fig. 8.9 **a** von Mises stress and **b** inelastic strain distribution in the critical solder joint at the end of impulse loading. The stress scale is in MPa

8.4.3 Stresses and Inelastic Strains in Critical Solder Joints

The predicted instantaneous distribution of von Mises stress in the critical solder joint, labeled E5 corresponding to time at the completion of the load pulse, is shown in Fig. 8.9a. The calculated maximum equivalent stress of 82.2 MPa is relatively high due to the high strain rate loading effect on the ductile bulk solder joint. The stress level for the solder corresponding to the strain of 0.32 pct. at a strain rate of 10^{-3} s^{-1} is in the order of 45 MPa (see Fig. 4.15). A high stress gradient develops at the edge of the solder/IMC interface on both device and board sides of the solder joint. This is likely due to geometrical stress concentration and local stiffness effects of different materials in the adjacent layers. It is worth noting that a relatively high stress region forms diagonally across the bulk solder. This indicates potential failure of the bulk solder by ductile shear mode under pulse loading. However, it is anticipated that fracture across the solder/IMC interface will dominate the failure process in view of the low strength of the brittle-like solder/IMC interface at high local strain rates due to impact loading.

The corresponding inelastic strain is confined only to a small edge region of the solder/IMC interface while the bulk solder remains elastic. The inelastic strain reaches a maximum of 0.32 pct. after 0.5 ms of the impulse has elapsed. Such localized inelastic strain could lead to crack initiation and dynamic crack propagation across the brittle solder/IMC interface following multiple board-level drop events.

The evolution of von Mises stress and the corresponding equivalent inelastic strains in the critical solder joint during the simulated drop pulse loading is illustrated in Fig. 8.10. The material point in the solder chosen for the history plots is located next to the solder/IMC interface on the device side of the assembly, marked *A* in Fig. 8.9a. The simulated pulse load is shown in Fig. 8.10a. The dynamic response of the solder to applied loading is expressed in terms of von Mises stress and inelastic strain, as shown in Fig. 8.10b, c, respectively. An initial delay in response of the material is due to the time required for the stress wave to travel from the loading (mounting) point of the test board to the solder in the BGA assembly. In fact, at the peak impulse load of 1500G (elapsed time of 0.25 ms), the von Mises stress in the critical solder only reached 18.4 MPa (or 22.3 pct. of the peak stress). A peak stress of 82.4 MPa is achieved at the end of the load pulse and then gradually decreases.

The corresponding inelastic strain initiated towards the end of the pulse after 0.45 ms has elapsed. The apparent high initial yield strength of the solder at 68 MPa indicates the high strain rate effect on the yield strength property. The average inelastic strain rate in the critical solder joint is 66.7 s^{-1}. In addition, inelastic strain continuously increases during the loading pulse as indicated by the continuous increase in the subsequent yield stress level. The largest magnitude of inelastic strain in the most critical solder joint at the end of the drop test is 0.375 pct.

8.4.4 Dynamic Damage Process of Solder/IMC Interface

The damage initiation variable, given by Eq. (7.6), is monitored at integration points of the cohesive solder/IMC interface elements throughout pulse loading. The distribution of the damage initiation variable in the solder/IMC interface region of the BGA solder joints following the board-level single drop test is illustrated in Fig. 8.11. Both interfaces on the device (top) and board (bottom) sides of the PCB assembly are shown for each solder joint. A damage initiation value in excess of 0.92 is arbitrarily taken to reflect the condition that damage has initiated at the material point. Results show that interface damage initiates and is localized in the outer peripheral solder joints followed by the inner peripheral solders

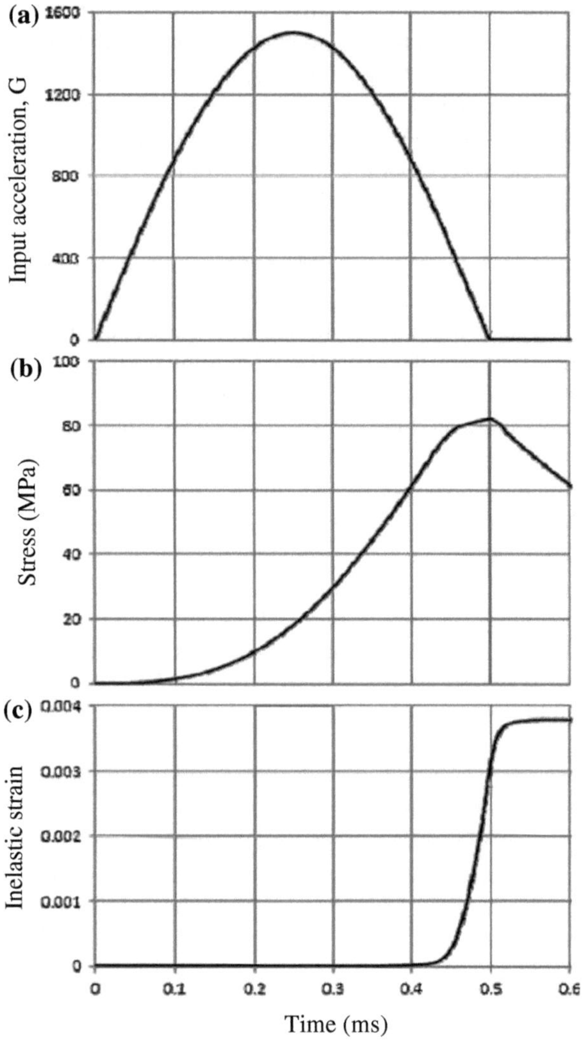

Fig. 8.10 **a** Input sinusoidal drop pulse loading, **b** evolution of von Mises stress and **c** equivalent inelastic strain in the critical solder

in the array. In these solder joints, damage initiated and progressed from the edge location closer to the center of the array for interfaces at the device side. Damage initiation occurs at opposite locations for interfaces at the board side of the package. This signifies the large induced shear stress across the bulk solder. Although damage initiated only on a small region of the solder/IMC interface, a fatigue crack could easily initiate at and propagate from this location during subsequent drop tests or reliability temperature cycles. Detailed results also indicate that the solder/IMC interface damage is greater at the board side of the BGA assembly.

Since the solder/IMC interface was simulated to behave as an ideally brittle interface in this case study, the region where interface damage initiated is

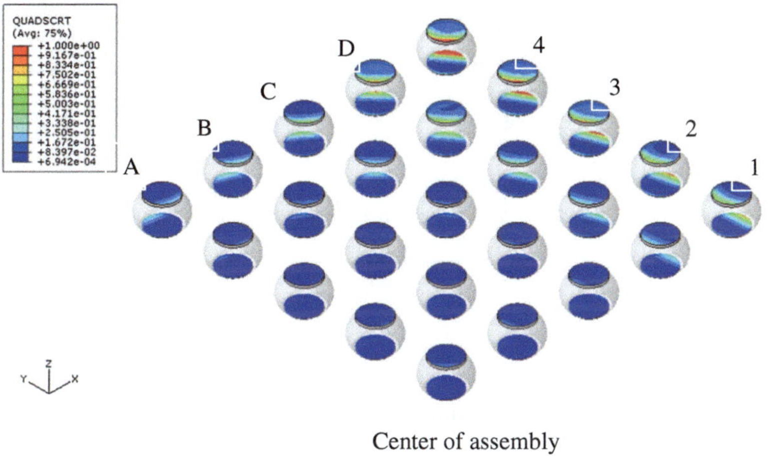

Fig. 8.11 Distribution of damage initiation variable, identified as *QUADSCRT* on solder/IMC interface at device (*top*) and board (*bottom*) sides of solder joints following the board-level drop test simulation

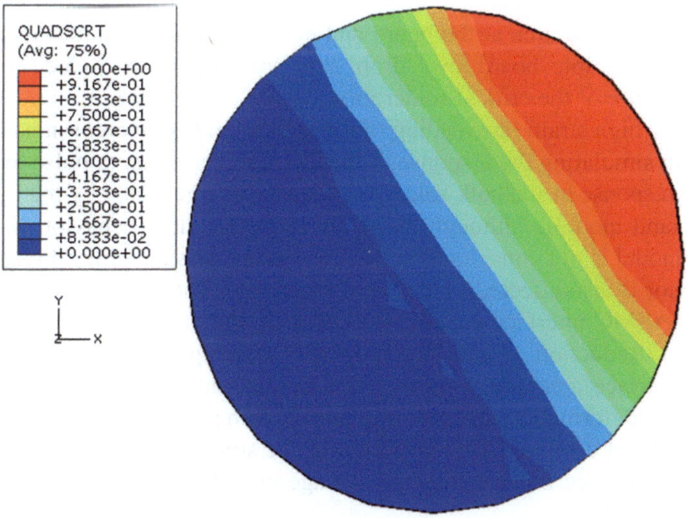

Fig. 8.12 Damaged (*cracked*) and non-damaged regions of the interface at the board side of the critical solder joint (E5) indicating a *straight crack* front

interpreted as interface crack. The predicted crack and non-crack regions of the interface in the critical solder joint E5 is shown in Fig. 8.12. Interface crack initiated at the edge location, primarily due to geometrical stress concentration factors, and rapidly progressed across the interface plane. The boundary between crack

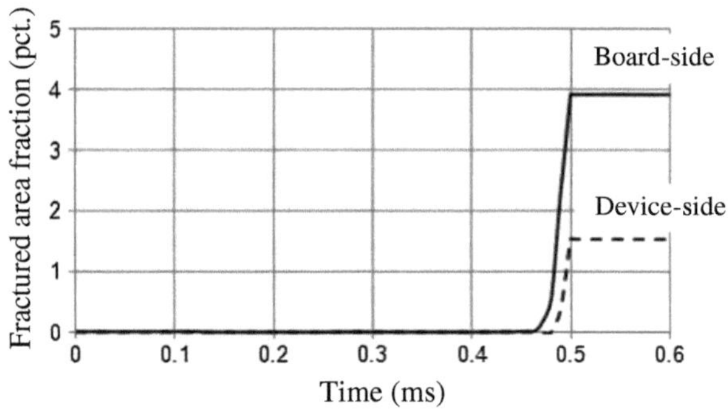

Fig. 8.13 Dynamic fracture process of solder/IMC interface at the board side (*solid line*) and device side (*dashed line*) of the critical solder joint during the single board-level drop test

(damage value >0.92) and non-damage, or an intact interface region, represents the interface crack front. The relatively straight dynamic crack front is attributed to the low toughness of the solder/IMC interface.

Rapid damage progression at the critical solder/IMC interface following the damage initiation event is described in terms of the evolution of the damage (fracture) area over the pulse time duration, as shown in Fig. 8.13. Damage progression occurs over the 0.02 ms of the impulse load indicating the dynamic fracture process. However, the fractured area of the interface is 3.92 and 1.55 pct. of the solder/IMC interface area at the board and device side of the solder joint, respectively. Thus, multiple board-level drop tests are required to induce a complete interface fracture of the critical solder joint.

Relatively high strain rate loading of the solder joint is inherent in a board-level drop test. In simulating the response of the BGA solder joints to pulse loading, the strain-rate response of the bulk solder is adequately represented by the constitutive model (Anand model). Although the strain rate-dependent behavior of the cohesive zone model is not considered in this case study, it is anticipated that the interface behavior is also affected by the local strain rate. The assumption of an ideally brittle solder/IMC interface tends to overestimate the prediction of interface crack propagation. Reduction of interface properties at a high strain rate would further facilitate interface cracking. However, the simulation appropriately describes the mechanics of deformation and the interface cracking process of solder joints during a board-level drop test. Consideration of the strain rate-dependent response of solder/IMC interface for the cohesive damage model is hindered by a lack of relevant measured data.

8.4.5 Aspects of FE Model Validation

The FE model for PCB assembly used in the simulation of a board-level drop test can be validated by comparing the predicted location of the critical solder joint with that observed in similar assembly. It was observed that corner solder joints incur more damage than those at other locations, while failed solder joints appear to gather along the diagonals of the array [9]. The simulated brittle solder/IMC interface fracture mechanism should also be validated. A board-level drop test for condition B has also demonstrated dominant cracking along solder/IMC interface [10]. In addition, similar solder/IMC interface fracture was observed at higher strain rates [3]. The interface fracture mode is defined when less that 10 pct. of the solder remained on the pad after the fracture event (e.g. [11]).

References

1. Albrecht, H.-J., Juritza, A., Müller, K., Müller, W.H., Sterthaus, J., Villain, J., et al.: Interface reactions in microelectronic solder joints and associated intermetallic compounds: an investigation of their mechanical properties using nanoindentation. In: 5th Electronics Packaging Technology Conference, pp. 726–731 (2003)
2. Anand, L.: Constitutive equations for hot-working of metals. Int. J. Plast. **1**(3), 213–231 (1985)
3. Darveaux, R., Reichman, C.: Ductile-to-brittle transition rate. In: 8th Electronics Packaging Technology Conference, 283–289 (2006)
4. JEDEC, JESD22-B111: Board Level Drop Test Method of Components for Handheld Electronic Products (2003)
5. JEDEC, JESD22-B104C: Mechanical Shock (2004)
6. Kim, I., Lee, S.-B.: Reliability and failure analysis of lead-free solder joints for PBGA package under a cyclic bending load. IEEE Trans. Compon. Packag. Technol. **31**(2), 478–484 (2008)
7. Lai, Z.B.: Classical and damage mechanics-based models for lead-free solder interconnect. Master Thesis, Universiti Teknologi Malaysia, Malaysia (2009)
8. Lai, Y.-S., Yang, P.-F., Yeh, C.-L.: Experimental studies of board-level reliability of chip-scale packages subjected to JEDEC drop test condition. Microelectron. Reliab. **46**(2–4), 645–650 (2006)
9. Lai, Y.-S., Yang, P.-C., Yeh, C.-L.: Effects of different drop test conditions on board-level reliability of chip-scale packages. Microelectron. Reliab. **48**(2), 274–281 (2008)
10. Liu, F., et al.: Board level drop test analysis based on modal test and simulation. J. Electron. Packag. **130**(2), 021007 (2008)
11. Park, T.-S.: A study on mechanical fatigue behaviors of ball grid array solder joints for electronic packaging. Korea Advanced Institute of Science and Technology (KAIST), Daejeon (2004)
12. Sano, M., Chou, C.-Y., Hung, T.-Y. and Yang, S.-Y.: Uncertainty and reliability analysis of chip scale package subjected to Board-level Drop test. 10th International Conference on Thermal, Mechanical and Multi-Physics Simulation and Experiments in Microelectronics and Microsystems (EuroSimE), 1–6 (2009)
13. Schubert, A., et al.,: Fatigue life models for SnAgCu and SnPb solder joints evaluated by experiments and simulation. In: Proceedings of 53rd Electronic Components and Technology Conference (2003)
14. Tong Yan, T. et al.: Novel numerical and experimental analysis of dynamic responses under board level drop test. Thermal and mechanical simulation and experiments in microelectronics and microsystems. In: Proceedings of the 5th International Conference on EuroSimE (2004)
15. Wong, E.H., Rajoo, R., Seah, S.K.W., Selvanayagam, C.S., van Driel, W.D., Caers, J.F.J.M., Zhao, X.J., Owens, N., Tan, L.C., Leoni, M., Eu, P.L., Lai, Y.-S., Yeh, C.-L.: Correlation studies for component level ball impact shear test and board level drop test. Microelectron. Reliab. **48**(7), 1069–1078 (2008)

Chapter 9
Application IV: Fatigue Fracture Process of Solder Joints

9.1 Introduction

An assessment of solder joint reliability for BGA packages and assemblies under cyclic mechanical loading was addressed in Chap. 6. Finite Element (FE) simulations of the test assembly with BGA solder joints subjected to cyclic flexural loading and torsional loading was described in Sects. 6.3 and 6.4, respectively. The simulations established stress-strain hysteresis behavior of the critical solder joint in the array. Phenomenological fatigue life models, as described in Sect. 4.6, were developed based on least-squared fit correlations between the calculated characteristic fatigue-life variables (such as the accumulated inelastic strain, $\varepsilon_{in,acc}$ and inelastic work density per load cycle, $W_{in,acc}$) and measured fatigue lives of identical assemblies. However, this classical life prediction method does not account for the observed multiple failure types occurring during the fatigue fracture process. In this respect, damage mechanics-based models, described in Chap. 7, are employed to simulate the dominant failure mechanisms observed during solder joint fatigue, namely ductile bulk solder failure and the relatively brittle solder/IMC interface fracture. The occurrence of a specific failure mode is dictated by the acting local strain rate, temperature and the material system that makes up the solder joint. Ductile fatigue failure across the bulk solder was observed under moderate strain rate loading (typically induced during temperature cycling), while a brittle fracture along the solder/IMC interface is common at high straining rates (such as that induced during the board-level drop test). Accurate reliability prediction of solder joints, therefore, should acknowledge the progressive damage process leading to fatigue failure of the assembly.

The fatigue failure process of solder joints constitutes of a crack initiation event followed by the crack propagation stage to the catastrophic fracture of the joints. This progressive failure process can be quantified and described using the damage mechanics approach. The cyclic damage process of the bulk solder joint can be expressed in terms of the accumulated inelastic hysteresis energy for both damage initiation and propagation criteria [2]. The characteristic evolution

M. N. Tamin and N. M. Shaffiar, *Solder Joint Reliability Assessment*, Advanced Structured Materials 37, DOI: 10.1007/978-3-319-00092-3_9,

of continuum damage in Pb-free solder joint under cyclic shear loading was examined using FE simulation [8]. The stabilized cyclic response of the solder can also be calculated directly from a converged solution using a combination of the Fourier series and time integration scheme [3]. In the relatively brittle solder/IMC interface, the cyclic damage process can be described based on the cohesive behavior of the interface, as discussed in Sect. 7.3. In the cohesive zone model (CZM), a stress-based criterion is prescribed for the damage initiation event while subsequent damage progression is predicted using an energy-based criterion. The CZM model, initially developed for monotonic loading, was modified to account for stress reversals and cyclic damage evolution laws, as discussed in Sect. 7.3.

This chapter addresses the mechanics of cyclic damage processes in solder joints when subjected to applied mechanical load cycles. Two case studies are presented; cyclic shear loading of a test assembly with a single solder joint and flexural loading of a test board with a BGA assembly. In the former, failure prediction of the solder joint is based on competing mechanisms of bulk solder failure and solder/IMC interface fracture in the solder joint. The response of the solder joint is quantified through FE simulation employing the cyclic continuum damage and cyclic cohesive zone model, as described in Chap. 7. The latter revisits the case study on cyclic flexural loading of the test assembly presented in Sect. 6.3, but with the addition of cyclic CZM to describe progressive fatigue failure of the solder/IMC interfaces. Details of FE simulation procedures, including geometry and element mesh, material properties and damage model parameters, cyclic loading parameters, and boundary conditions are described. The FE simulation results are discussed in terms of inelastic strain, stress, and damage evolution in the bulk solder and solder/IMC interface.

9.2 Cyclic Shear Loading of a Solder Joint

In view of the complexity of performing a cyclic shear test on a microelectronic assembly with BGA solder joints, a test assembly with a single solder is considered. The objective of the test is to assess the reliability of the solder joint through an understanding of the deformation response and failure process of the solder under shear-dominant load cycles.

9.2.1 Mathematical Model

A test assembly with a single solder joint is considered in this case study. The symmetry of the geometry, loading and boundary conditions in the *xz*-plane enable one-half section of the test assembly to be considered for the simulation, as shown in Fig. 9.1. The assembly consists of a single SAC405 solder joint with a

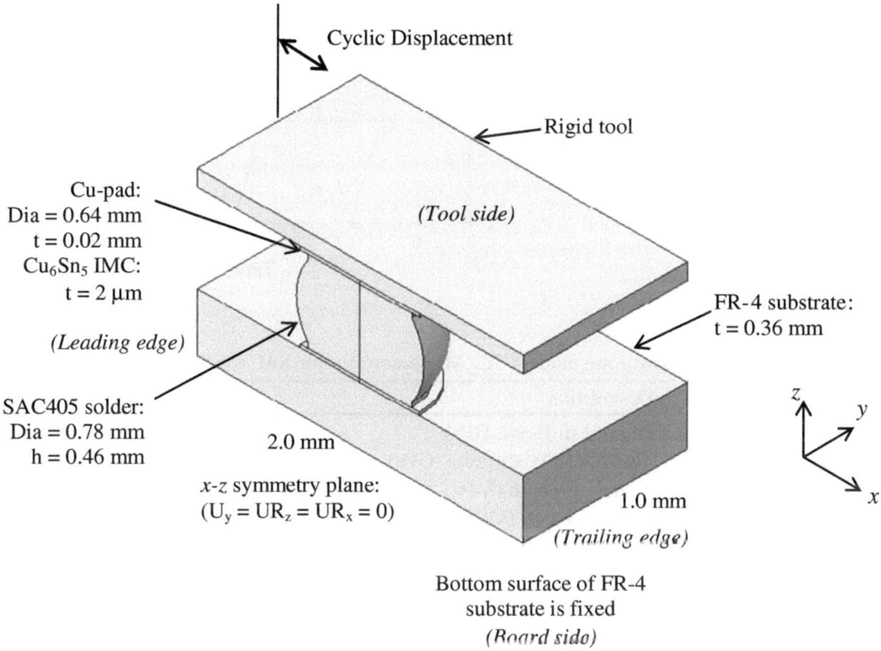

Fig. 9.1 Mathematical model of the test assembly with single solder joint

diameter and a stand-off height of 0.78 and 0.46 mm, respectively. Copper pads of diameter 0.64 mm and thickness 0.02 mm are modeled between the solder and the *rigid* tool at the top side and the 0.36 mm-thick FR-4 substrate at the bottom side of the assembly. A stable thin layer of Cu_6Sn_5 IMC is assumed to have formed between the solder and Cu pad. The rigid tool represents the stiff Si-die in a flip chip package.

The symmetry boundary conditions are prescribed as:

Symmetry about *xz*-plane: Displacement, $U_y = 0$ and rotations, $U_{Rx} = U_{Rz} = 0$

The bottom surface of the FR4 substrate is constrained from displacement and rotation.

Cyclic shear loading is introduced by applying displacement cycles to the leading edge of the rigid tool. Thus, the tool oscillates only in the horizontal *x*-direction. The prescribed displacement cycle consists of a displacement range of 0.003 mm, mean displacement of 0.0015 mm and displacement ramp rate of 0.006 mm/s, as illustrated in Fig. 9.2. The simulation is performed cycle-by-cycle with a total of 50 displacement cycles.

The cyclic shear test of the assembly is simulated at an ambient temperature of 30 °C. The strain rate-dependent response of the bulk SAC405 solder joint is represented by the Anand model (described in Sect. 4.5). Material parameter values for the model along with the optimized set of modified Anand model parameters

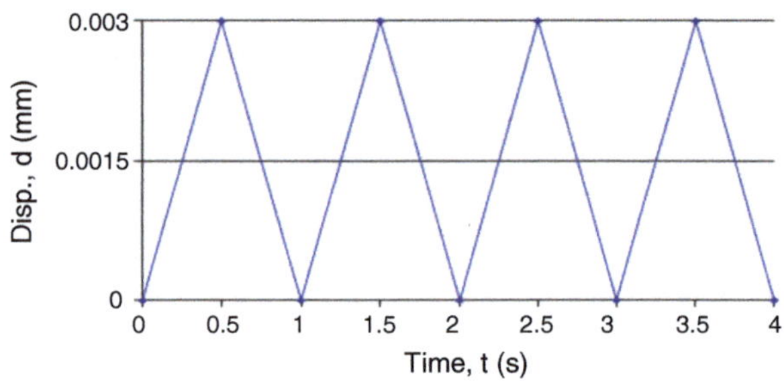

Fig. 9.2 Cyclic displacement loading on the test assembly with single solder joint. Displacement range of 0.003 mm and load ratio, $R = 0$

Table 9.1 Cyclic cohesive zone model (CCZM) parameters for $SAC405/Cu_6Sn_5$ interface

Parameter	Description	Value
$k_{n,0}$	Penalty stiffness (GPa)	41
$k_{s,0}$	Shear penalty stiffness (GPa)	15
T_0	Tensile strength (MPa)	56.77
S_0	Shear strength (MPa)	43.66
$G_{IC,0}$	Critical energy release rate for normal mode (N/mm)	0.04
$G_{IIC,0}$, $G_{IIIC,0}$	Critical energy release rate for shear mode (N/mm)	0.065
η	Mixed-mode fracture parameter	1
b	Cyclic damage exponent	−0.175

are identical to that shown in Table 8.1a. Properties for other materials used in the test assembly are summarized in Table 8.1b.

The cyclic damage response of the solder/IMC interface is simulated using the cyclic cohesive zone model (CCZM), as described in Sect. 7.3. The characteristic degradation of CCZM parameters for the $SAC405/Cu_6Sn_5$ interface is expressed in power-law form with the applied load cycles. The coefficient of the degradation expression represents the initial value of each model parameter, as listed in Table 9.1. These values also correspond to those for monotonic loading ($N = 0.5$ cycle), as summarized in Table 8.1c. In the absence of measured fatigue data for a $SAC405/Cu_6Sn_5$ interface, the cyclic damage exponent, $b = -0.175$ was determined from limited shear fatigue test data on a SAC405 solder alloy [4]. An example of cyclic degradation of tensile and shear strength of $SAC405/Cu_6Sn_5$ interface, based on initial values given in Table 9.1, is illustrated in Fig. 9.3.

The bulk solder joint degrades under fatigue loading as predicted by the cyclic continuum damage model. The damage model, as described in Sect. 7.4, is based on matching the accumulated inelastic hysteresis energy per stabilized cycle, as predicted by the FE simulation with measured fatigue crack growth rate data of the solder joint. Cyclic continuum damage model parameters for a SAC405 solder are shown in Table 7.4. A variable for each damage model is monitored throughout the loading cycles. Once a particular damage mode has initiated, property degradation of the material point follows, thus causes a subsequent response of the damaging material.

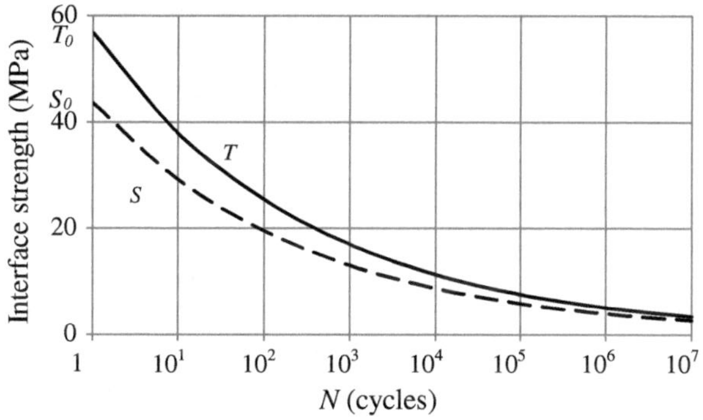

Fig. 9.3 Characteristic degradation of fatigue properties (tensile and shear strength) for the cohesive SAC405/Cu_6Sn_5 interface based on Eq. (7.15)

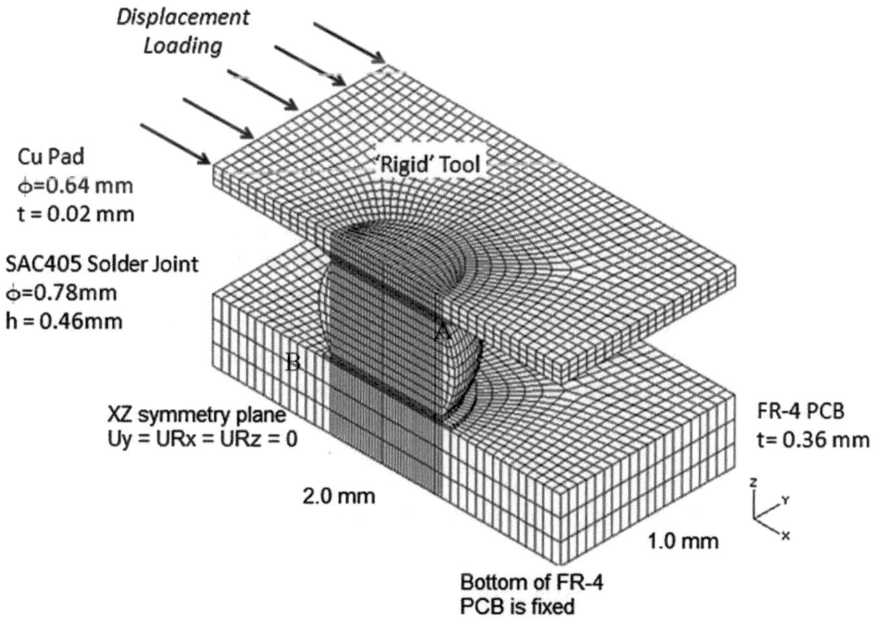

Fig. 9.4 FE model of the test assembly with final mesh density showing relative size of elements in various components of the assembly

9.3 Finite Element Modeling

The geometry of the test assembly with the single solder joint is discretized using 8-node continuum elements. Figure 9.4 shows the final element mesh density for the various parts of the test assembly. A layer of *COHESIVE ELEMENTS* with *zero* thickness is placed between the solder joint and the IMC layer on the tool and board side of the assembly. Such element will carry normal and shear tractions on its surface.

A mesh convergence study was performed on the model to determine the largest size of elements in the bulk solder that would ensure mesh-independent results. In this analysis, the element size for the solder, particularly in the vicinity of the solder/IMC interface, is successively reduced while the calculated stress at the selected critical material point is monitored. The calculated stress fields in the solder joint for successive simulation with decreasing size of the element are compared in Fig. 9.5a. Identical stress contours in the last two successive simulations are established with an element size of 0.015 mm or smaller in the vicinity of the solder/IMC interface region of the converged element mesh. A variation of von Mises stress at a selected critical location in the solder with the size of the element, as shown in Fig. 9.4b, conveys similar information on the converged element mesh. A subsequent analysis is performed using element mesh with an element edge length of 0.01 mm for regions in the vicinity of the solder/IMC interface. The adjacent cohesive interface element has matching nodes with that of the elements in the bulk solder joint.

9.4 Stress and Inelastic Strain Distribution in the Solder Joint

The results of the simulation are employed to provide a quantitative description of the progressive material damage in the SAC405 solder and/or SAC405/Cu_6Sn_5 interface during the cyclic shear-dominated failure process. The FE results are interpreted and discussed in terms of the characteristic evolution of inelastic strain, stress and damage variable to failure, instantaneous stress and inelastic strain distribution and stress-strain hysteresis of the solder material.

9.4.1 Stress Distribution in the Solder Joint

Distributions of equivalent (von Mises) stress in the bulk solder joint and on the solder/IMC interface plane corresponding to selected displacement cycles are shown in Fig. 9.6. Each stress field corresponds to the peak displacement for the selected cycle, as indicated in the figure. The highest stress magnitude and gradient are initially predicted to occur at both leading and trailing edges of the solder/IMC interface on the tool side of the assembly. These are likely due to the effect of local geometry (stress raiser) and mismatches in the modulus of the SAC405 solder and the adjacent Cu_6Sn_5 IMC layer. Such excessive stress causes interface damage to initiate during the first cycle (as discussed later using Fig. 9.9b). Force equilibrium leads to a continuous decrease of the equivalent stress magnitude in the solder joint with the applied displacement cycles. Following the interface crack initiation that occurs during the 18th cycle (refer to Fig. 9.9b), the peak stress shifted to the interface crack front while the stress on the separated interfaces diminished,

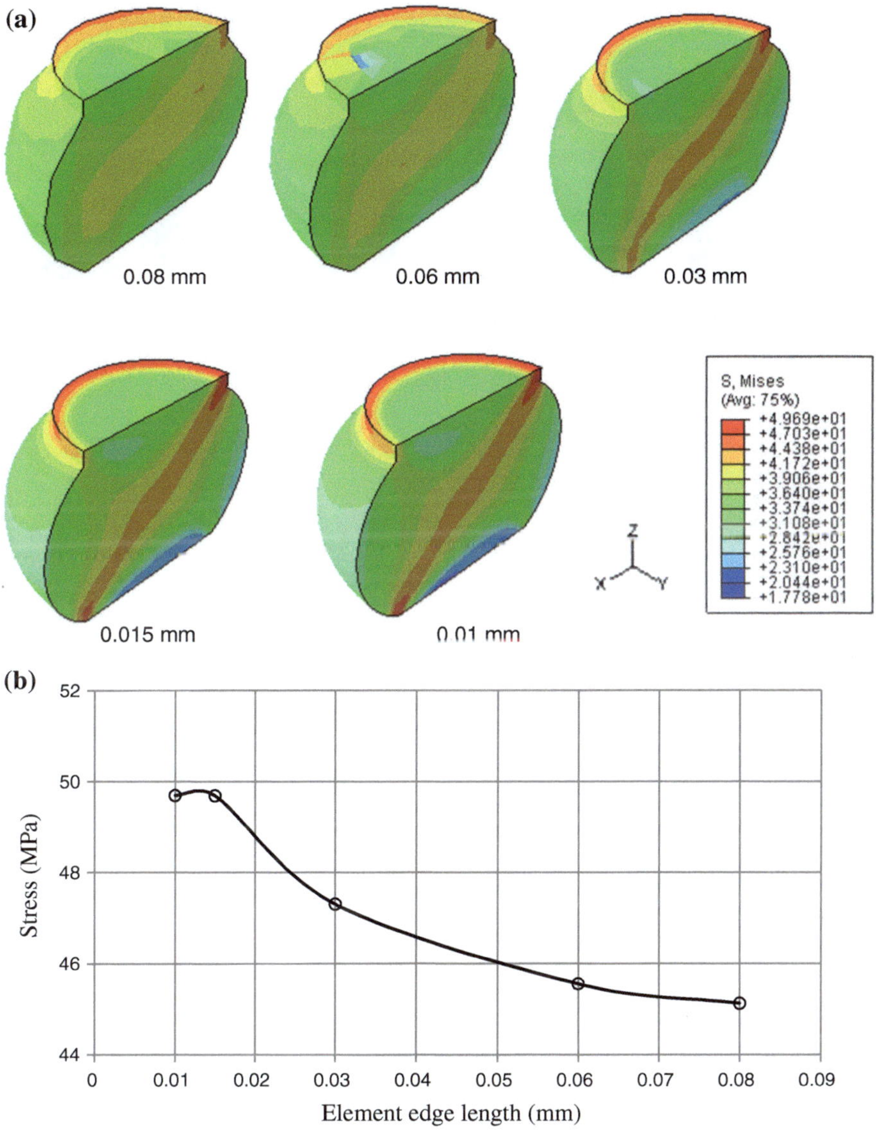

Fig. 9.5 **a** Comparison of the calculated stress fields in the solder joint when different size elements are used. **b** Variation of the predicted von Mises stress at a critical point in the solder with different size elements. The converged element size has an edge length of 0.015 mm

as illustrated by the stress contour corresponding to $N = 24.5$ cycle. In this discussion, the crack front is defined by the boundary between the bonded and debonded region of the interface. Stress redistributes as the traction-free interface crack surface propagates during subsequent cycles. Similar solder/IMC interface fatigue crack was also observed to occur at the device or the tool side of the BGA package under mechanical loading [1, 6].

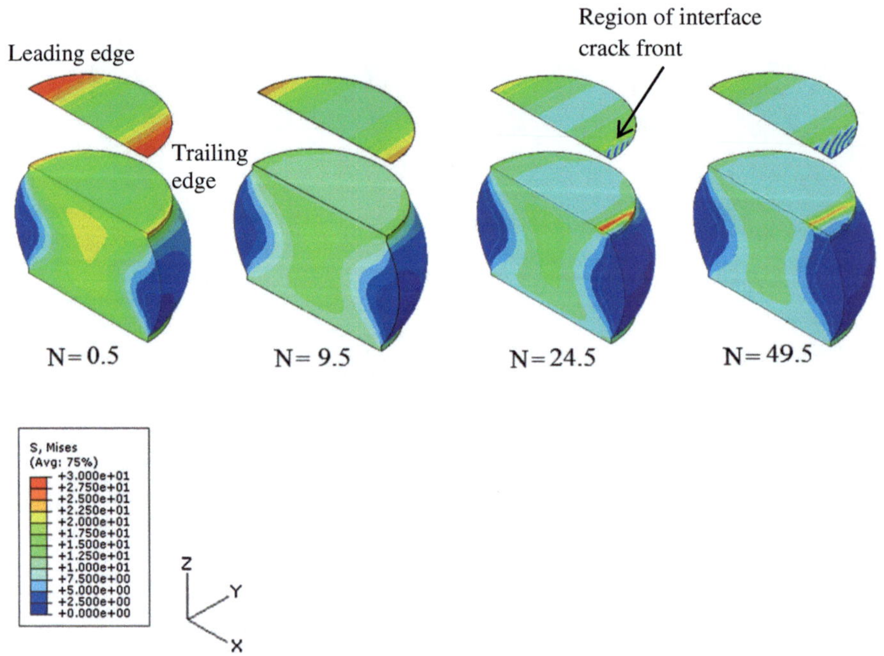

Fig. 9.6 Instantaneous stress distribution in damaging solder and solder/IMC interface material at selected accumulated displacement cycles. The stress scale is in MPa

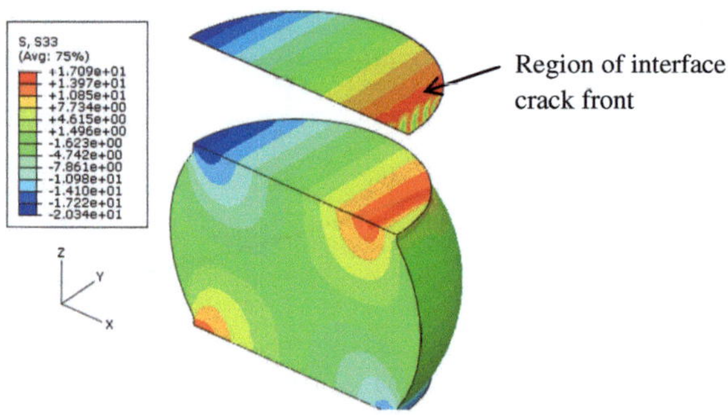

Fig. 9.7 Bending stress distribution in the solder joint corresponding to the peak of the accumulated displacement cycle ($N = 25.5$ cycles)

It is worth examining the induced bending effect due to the stand-off height of the solder joint under the applied shear loading. This effect is shown in Fig. 9.7 in terms of the normal stress component, σ_z and corresponds to the peak of the endured displacement cycle at $N = 25$ cycle. A bending effect manifests in tensile and compressive stresses at the trailing and leading edge regions of the solder joint near the rigid tool side of the test assembly, respectively. The tensile bending stress facilitates damage initiation under the opening Mode-I loading at the trailing edge of the solder/IMC interface. The high tensile stress and stress gradient are associated with the interface crack front. The practically straight contour of the localized stress suggests a straight interface crack front that reflects the brittle-like nature of the solder/IMC interface fracture process. Anti-symmetric distribution of stresses with respect to the board (bottom) side of the solder joint suggests potentially high shear stress along the diagonal plane of the solder joint.

9.4.2 Accumulated Inelastic Strain in the Solder Joint

The distribution of the accumulated equivalent inelastic strain in the bulk solder joint at the end of the applied displacement cycles ($N = 50$ cycles) is shown in Fig. 9.8a. Fracture of the solder joint is predicted to occur due to interface fatigue crack propagation beginning during cycle No. 18. Consequently, the accumulated inelastic strain is relatively low at 0.027 pct. and confined to the region in the vicinity of the interface crack front.

The evolution of inelastic strain during fatigue loading for three different material points in the solder joint is compared in Fig. 9.8b. Inelastic strain evolves early at the critical edge locations along the symmetry plane of the FE model (point marked *A* and *B*). A higher rate of strain evolution is demonstrated by the material point that experienced tensile stress (at *A*) when compared to that under compressive stress (at *B*). Inelastic strain at point *A* saturates at a low level of 4.1×10^{-8}. At $N = 18$ cycle, solder/IMC interface crack occurs resulting in a severe geometrical stress concentration point at the edge of the crack front for the location marked *C*. Consequently, inelastic strain begins to accumulate at a relatively faster rate before saturating at the greater magnitude of 0.027 pct. Saturation of the accumulated inelastic strain in the solder is likely due to the creation of traction-free interface crack surfaces in the adjacent region.

9.5 Characteristic Damage Evolution

In this case study, in the applied displacement cycles with a range of 0.003, $R = 0$ only results in solder/IMC interface damage and fracture. Damage of the bulk solder was not predicted during the first 50 cycles of the applied displacement. The progression of cyclic damage at the interface is represented by the contour of the damage variable, *SDV1* for selected displacement cycles, as shown in Fig. 9.9a. A damage variable magnitude of *SDV1* > 0.917 is taken to indicate the onset of separation of the adjacent material points at the interface. Results show that solder/IMC interface damage under cyclic shear loading initiates from the edge of the interface and propagates inwards on the interface plane on the device side of the assembly. At a sufficiently large number of enduring load cycles, a new

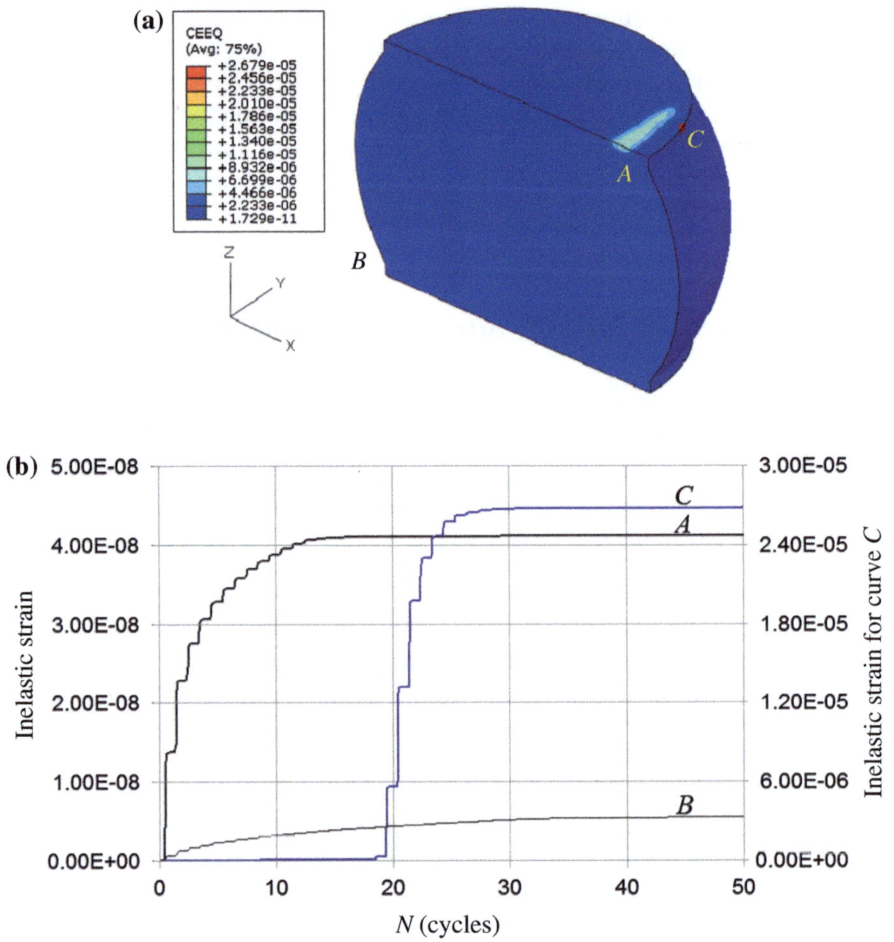

Fig. 9.8 **a** Inelastic strain distribution in the solder joint at $N = 50$ cycle. **b** Evolution of inelastic strain at selected material points in the solder joints during the first 50 cycles (displacement range of 0.003 mm, $R = 0$). Note the different strain scale used for curve C

interface crack nucleated on the opposite edge, but on the board side of the assembly. The induced shear stress across the diagonal of the bulk solder joint could potentially cause the fatigue crack to traverse across the bulk solder.

The evolution of surface tractions (normal and shear) at the critical point on the trailing edge of the solder/IMC interface during the interface cracking process is shown in Fig. 9.9b. The corresponding evolution of interface damage is also superimposed on the plot. Results indicate that both normal and shear tractions on the solder/IMC interface increase linearly during the first load cycle. However, the stresses decrease continuously in subsequent fatigue cycles due to the cycle-dependent degradation of strength properties. Such degradation is quantified by the simultaneous increase in magnitude of the interface damage variable. The

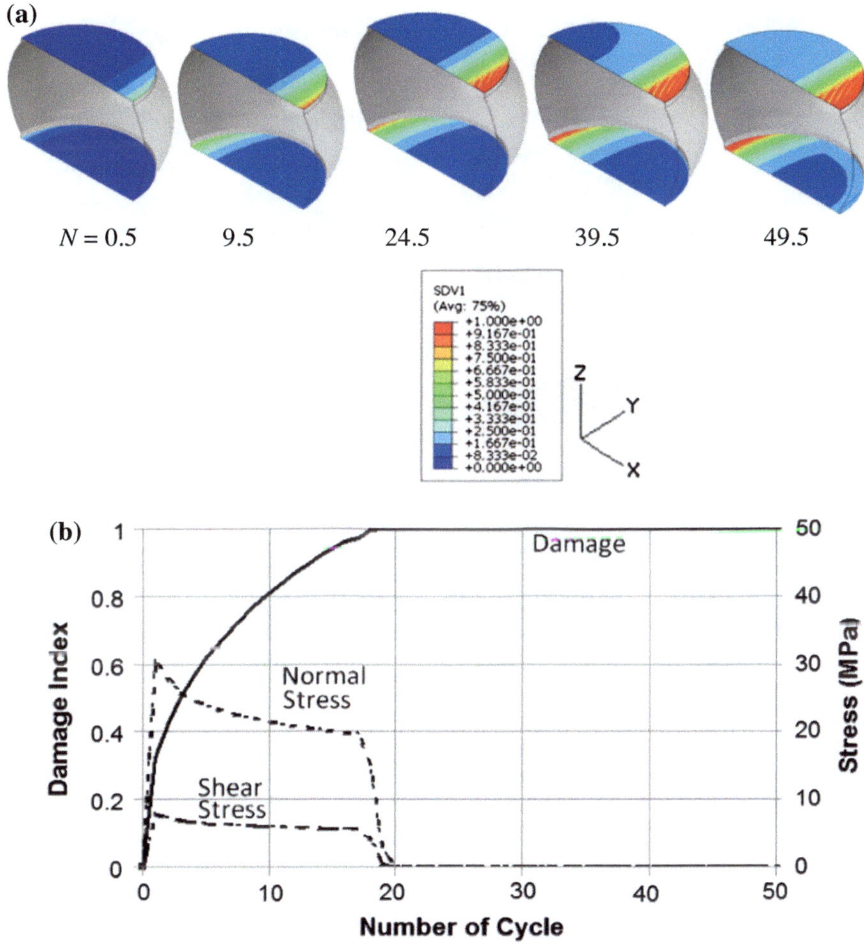

Fig. 9.9 **a** Progressive accumulation of solder/IMC interface damage at tool and board sides of the solder joint. **b** Evolution of stresses and damage propagation variable at the critical point of solder/IMC interface during fatigue cycles

damage index reaches the value for damage initiation at $N = 18$ cycles. Since the SAC405/Cu_6Sn_5 interface is assumed in this case study to be ideally brittle, damage initiation and interface separation are predicted to occur simultaneously. The tractions rapidly diminish for the separated interface material point. Collection of separated material points is interpreted in this case study to form the traction-free interface crack surfaces. An equilibrium of forces will cause the neighboring material points to endure greater tractions.

In the interest of illustrating the simultaneous occurrence of solder/IMC interface and bulk solder damage mechanisms, a new fatigue load case is examined. A greater magnitude of the applied displacement, similar to that shown in Fig. 9.2

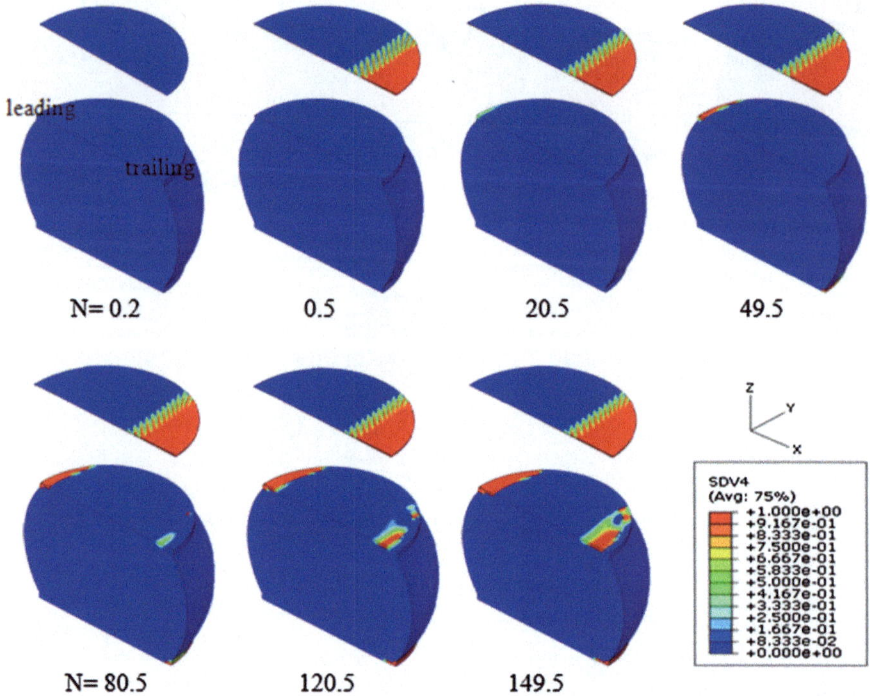

Fig. 9.10 Distribution of the accumulated damage propagation variable in the bulk solder joint and the cohesive solder/IMC interface. Separation is considered to occur for damage value, $D > 0.917$, as represented by *red* region. Displacement range is 0.01 mm, $R = 0$

but with a range of 0.01 mm, $R = 0$ is employed. The resulting progression of cohesive damage in the solder/IMC interface and continuum damage of the bulk SAC405 solder joint for selected displacement cycles is shown in Fig. 9.10. Damage variable magnitude *SDV4* > 0.917 denotes the separation of a material point while a lower damage value is associated with the degradation of the material stiffness property. Results show that solder/IMC interface separation or crack dominates the fatigue fracture process. An interface crack propagated to a significant length in the first half of the displacement cycle. After accumulating 50 fatigue cycles, a new crack initiates in the leading edge of the bulk solder in the vicinity of the solder/IMC interface. This crack then propagates at a slow rate in the bulk solder, along the interface plane with subsequent fatigue cycles while the interface crack practically stalled. The competition between bulk solder failure and solder/IMC interface cracking is influenced by the local strain rate acting at each material point, as it is likely to govern stiffness and fracture strength of the materials. The strength of the ductile solder increases while the relatively brittle fracture strength of the solder/IMC interface decreases with increasing strain rates. However, strain rate-dependent properties and CZM parameters of

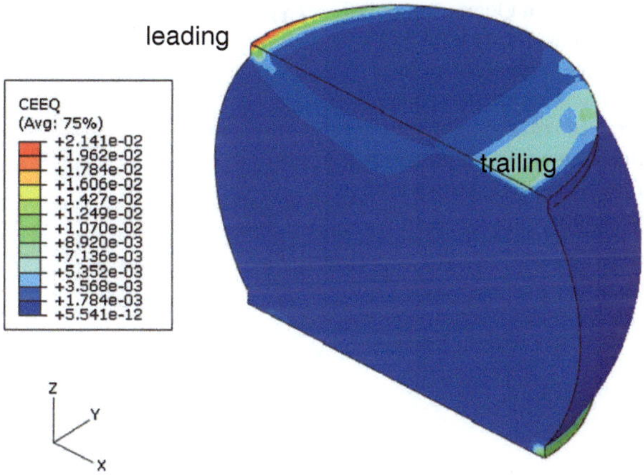

Fig. 9.11 Inelastic strain distribution in the solder joint corresponding to the peak of the first displacement cycle. Displacement range is 0.01 mm, $R = 0$

the SC405/Cu_6Sn_5 interface is not considered in the current simulation, while the rate-dependent response of the bulk SAC405 solder joint is well represented by the Anand model.

The distribution of inelastic strain in the damaging bulk solder joint corresponding to the peak of the first displacement cycle is shown in Fig. 9.11. Extensive inelastic strain is predicted along the leading and trailing edge region of the solder near the solder/pad interface for both board and tool sides of the test assembly. Plastic strain accumulation ceases at the trailing edge following interface separation where the stress diminishes. However, the intact leading edge continuously sustains the high stress that causes greater inelastic strain, as shown in the figure. Such high magnitude of inelastic strain, in the order of 2 pct., during the first one-half of the fatigue cycle is due to the large applied displacement range of 0.01 mm of the fatigue cycles (load ratio, $R = 0$). Although the region of extensive plastic strain is small, fatigue damage is likely to initiate leading to crack propagation from these locations in the solder joint. Failure of a solder joint by fatigue crack propagation is accompanied by limited plasticity associated with the crack front while the remaining bulk solder joint is elastic. Although a similar distribution of inelastic strain with damage in the bulk solder joint is displayed, a correlation of the characteristic evolution of inelastic strain with continuum damage during fatigue loading could not be ensured. Thus, inelastic strain could not explicitly represent the evolving state of damage in the solder joint.

9.6 Cyclic Flexural Loading of BGA Assembly with Damaging Solder Joints

An FE simulation of a test assembly with BGA solder joints subjected to cyclic flexural loading was presented and discussed in Sect. 6.3. The fatigue life of the critical solder joint in the assembly was predicted based on characteristic fatigue

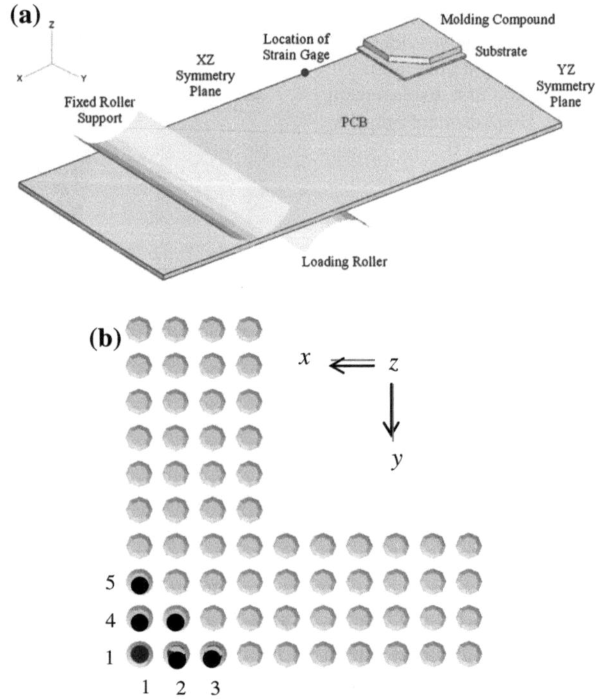

Fig. 9.12 **a** Quarter model of the test assembly with BGA package in the four-point bend test setup. **b** Peripheral array of solder joints with fine element mesh prescribed for solder joints shown with dark spot

parameters of the bulk solder, namely the accumulated inelastic strain per cycle, $\varepsilon_{in,acc}$, and inelastic work density per cycle, $W_{in,acc}$. Such a phenomenological prediction method relies on pre-established correlations of the characteristic fatigue parameters of the solder joint with measured fatigue lives of test assemblies with BGA solder joints. In this chapter, the previous case study discussed in Sect. 6.3 is re-examined to illustrate a prediction of the failure process employing a continuum damage-based fatigue model for bulk solder joints. This simulation approach enables the prediction of the damage initiation event, subsequent crack (material separation) propagation in the bulk solder joint and the path for crack propagation.

9.6.1 Finite Element Modeling

The geometry of the test assembly consisting of a test board with a surface mounted BGA package, as described in Fig. 6.8a, is reproduced in Fig. 9.12a for convenience. Identical symmetry boundary conditions of the quarter model of the test assembly are considered. The simulation employs Load case LC1 that constitutes of sinusoidal applied load to the test assembly under the four-point bend test set-up. The load case LC1 consists of the load range $\Delta P = 25$ N and load ratio $R = 0.0385$.

A strain rate-dependent response of the bulk solder joint is represented by the Anand model with refined model parameters for SAC405 solder joints. The Anand model parameter values for SAC405 solder joint are listed in Table 4.5 (Chap. 4). Properties of other materials of the test assembly are shown in Table 6.3 (Chap. 6). The cyclic continuum damage model described in Sect. 7.4 with model parameter values given in Table 7.4 (Chap. 7) is employed to describe the damage and failure process of the bulk solder joint. Solder/IMC interfaces are assumed to remain intact throughout the flexural load cycles.

Based on a mesh convergence study for the quarter model of the test assembly, the solder joint located at the corner of the peripheral array is the most critical solder joint, as illustrated in Fig. 6.8b. This solder joint was discretized into fine element mesh, as shown in Fig. 6.10. In view of the potential damage initiation and progression into the neighboring solder joints, additional solder joints in close vicinity of the corner solder are discretized into identical high mesh density, as indicated by dark solders in Fig. 9.12b.

9.6.2 Distribution of Internal States in the Critical Solder Joint

Internal states of displacement, strain and stress in damaged solder joints are calculated at every load increment throughout the loading cycles. Distribution of these variables in the critical corner solder joint after enduring 80 flexural load cycles is shown in Fig. 9.13a through 9.13c. The illustrated conditions correspond to the peak of the load cycle ($N_f = 80$ cycles). The boundary between separated ($D = 1$) and intact material points is interpreted as the crack front while the separated material points or elements represent the traction-free crack surfaces, as illustrated in Fig. 9.13c. It is noted that this solder crack propagates almost parallel to the solder/IMC interface due to high stress and a stress gradient in the solder in the vicinity of the assumed non-damaging interface. The relatively straight crack front suggests a brittle-like fracture. The strong influence of a local strain rate manifests in the high magnitude of stress in the order of 71 MPa in the vicinity of the crack front, as shown in Fig. 9.13a. The strain rate at this locality could be inferred from the slope of an inelastic strain curve at 80 cycles, as shown in Fig. 9.14. Equilibrium of forces across the traction-free crack surfaces decelerates the growth of the main crack while a secondary crack forms at the opposite edge of the solder with subsequent load cycles. The equivalent or von Mises stress concentrates along the crack front and diminishes over the wake of the crack surfaces. Figure 9.13b shows the corresponding equivalent inelastic strain in the critical solder joint. Inelastic strain accumulates with the evolving damage and confines to the region of the crack surfaces. The remaining bulk solder joint remains elastic. A similar contour of damage (or crack) to that of inelastic strain distribution suggests that solder joint fatigue failure is dictated by the accumulated inelastic strain in the solder joint. It is worth mentioning that

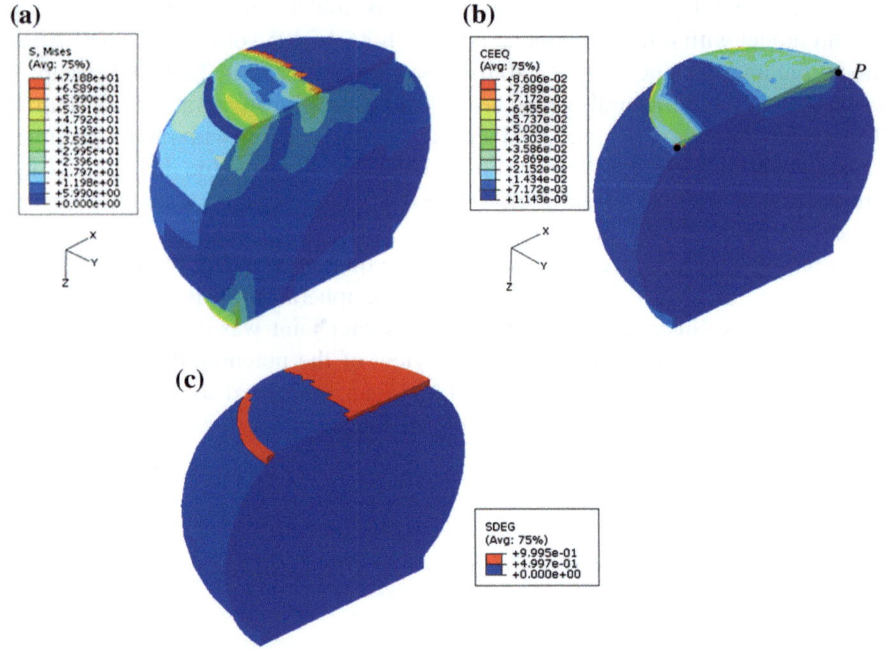

Fig. 9.13 Distribution of variables in the critical solder joint after accumulated 80 flexural load cycles ($\Delta P = 25$ N, $R = 0.0385$); **a** von Mises stress, **b** equivalent inelastic strain and **c** material separation variable representing crack surfaces (shown by *red* colored region)

the predicted fatigue damage and crack propagation in the region of solder/IMC interface on the board side of the assembly has also been reported experimentally in similar BGA packages and loading [5].

9.6.3 Evolution of Inelastic Strain and Propagation of Damage in Solder Joints

The characteristic evolution of equivalent inelastic strain in the corner solder joint at the critical material point marked *P* is shown in Fig. 9.14a over the applied flexural load cycles. The evolution of inelastic strain in neighboring solder joints is compared, as shown in the figure. During the first 50 load cycles, inelastic strain accumulates at a slow rate until the initiation of a small separated (crack) region along the edge of the solder next to the solder/IMC interface (0-A). This is followed by a sudden increase of inelastic strain to a rate of 0.33 pct. per cycle (0.33 pct s^{-1} at 1 Hz) during the crack propagation stage (A–B–C) for an additional 40 flexural load cycles. A short transition stage (C–D) signified by a slow inelastic strain rate is predicted, likely due to effects of load shedding to

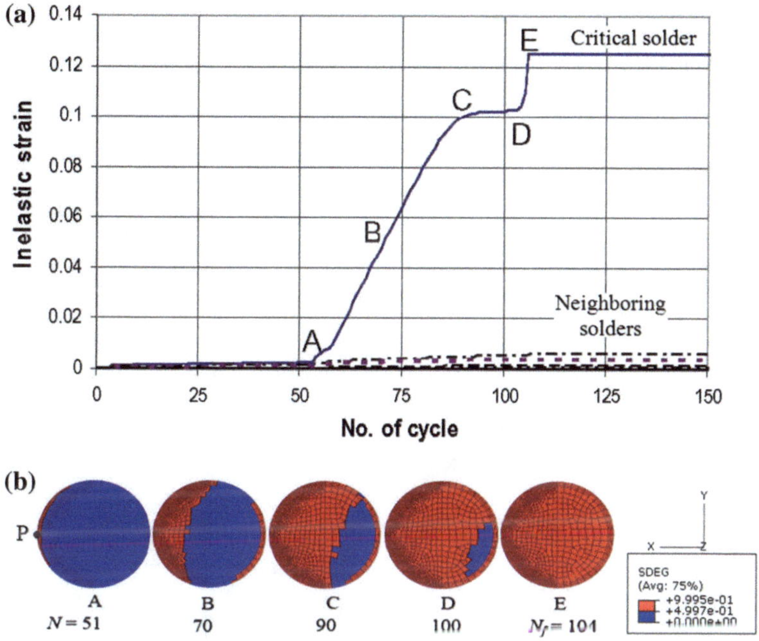

Fig. 9.14 **a** Characteristic evolution of equivalent inelastic strain in the critical and neighboring solder joints. **b** Crack propagation in the critical corner solder joint at selected elapsed flexural load cycles. *Red* colored region represents fractured surfaces

neighboring intact solder joints. The final stage of unstable fracture of the critical solder joint (D–E) occurred at an inelastic strain rate of 1.53 pct s^{-1}.

The reliability of the assembly at the prescribed flexural load cycles is quantified by the fatigue life of the critical solder, $N_f = 104$ cycles. It is worth noting that the damage initiation life covers nearly half of the total fatigue lives of the BGA package. Similar evolution characteristics, but at much slower inelastic strain rate, are demonstrated for neighboring solder joints, even after the critical corner solder joint has completely fractured (point E).

We recall the fatigue model based on the accumulated inelastic strain per load cycle for similar BGA assembly and identical reliability test setup shown in Fig. 6.15 [7]. The predicted life of the critical solder joint based on this classical mechanics model is $N_f = 410$ cycles. It is argued that the damage-based model demonstrated in this chapter acknowledges both the fatigue crack initiation stage (0-A) and the stable crack propagation stage (A-B-C) up to the final fast fracture of the critical solder joint. Once damage initiated at a material point in the solder, the stress is modified due to the continuous cyclic degradation of the elastic modulus. The separation of the material point further redistributes the load in the locality resulting in a stress gradient that accelerates the fracture process. Thus, the damage mechanics model accounts for the fatigue damage mechanisms of the solder joint.

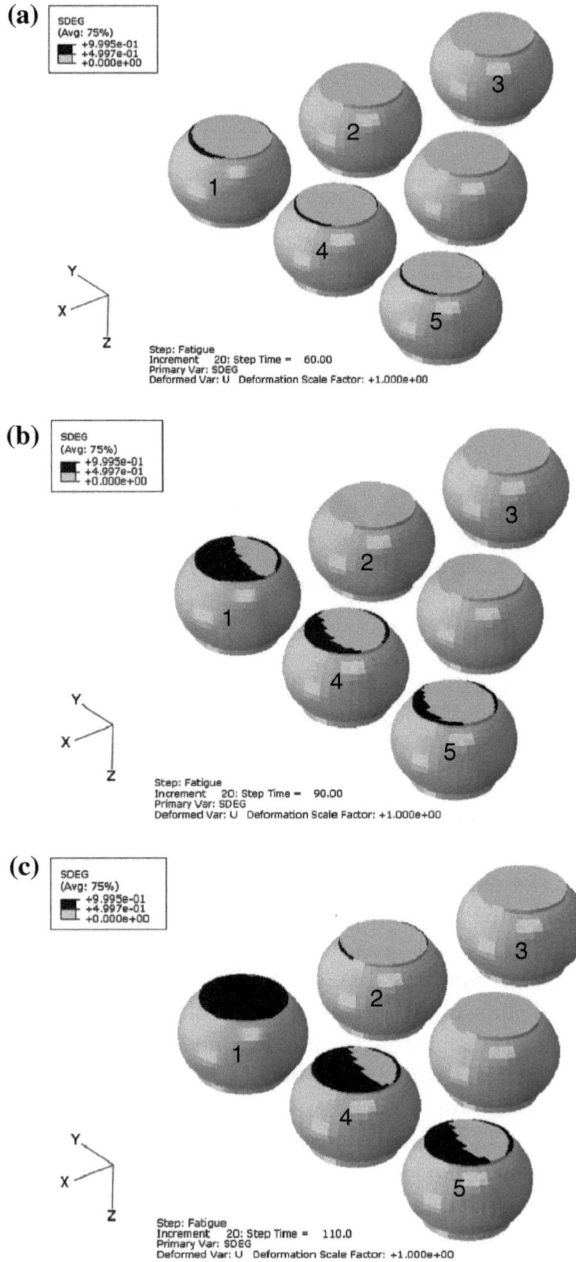

Fig. 9.15 Propagation of fractured region (crack) in BGA solder joints subjected to cyclic flexural load ($\Delta P = 25$ N, $R = 0.0385$); **a** after enduring 60 cycles, **b** 90 cycles and **c** 110 cycles

Distribution of the separated (crack) region in selected solder joints of the array after accumulated a given number of applied flexural load cycles is shown in Fig. 9.15. Solder joints labeled 1, 4 and 5 lie along the width direction (y-axis). Results show that damage and crack initiate at the edge of the solder/IMC interface for solder joints located along the outer peripheral array and across the width of the PCB, beginning with the corner solder joint, as shown in Fig. 9.15a, after 60 cycles. These solder joints experience the greatest relative shear deformation between the substrate and PCB throughout the flexural load cycles. Crack grows fastest in the corner solder joint (No. 1) followed by the neighboring solders (No. 4 and No. 5), as depicted in Fig. 9.15b, after accumulating 90 cycles. The crack propagated through the entire critical section in the corner solder, parallel to the solder/IMC interface plane, while covering greater than 50 pct. of the section near the solder/IMC interface of the neighboring solder, No. 4 (Fig. 9.15c). Meanwhile, an edge crack initiated in the neighboring solder, No. 2. The predicted results compares well with that observed experimentally [5].

References

1. Caers, J.F.J.M., et al.: A study of crack propagation in lead-free solder joints under drop impact. Electronic Components and Technology Conference. pp. 1166–1172 (2008)
2. Darveaux, R.: Effect of simulation methodology on solder joint crack growth correlation and fatigue life prediction. J. Electron. Packag. **124**, 147–154 (2002)
3. Dassault systèmes: abaqus documentation, Version 6.9 (2009)
4. Khatibi, G., et al.: A novel accelerated test technique for assessment of mechanical reliability of solder interconnects. Microelectron. Reliab. **49**(9–11), 1283–1287 (2009)
5. Kim, I., Lee, S.-B.: Reliability and failure analysis of lead-free solder joints for PBGA package under a cyclic bending load. IEEE Trans. Compon. Packag. Technol. **31**, 478–484 (2008)
6. Kim, K.S., Huh, S.H., Suganuma, K.: Effects of intermetallic compounds on properties of Sn–Ag–Cu lead-free soldered joints. J. Alloy. Compd. **352**, 226–236 (2002)
7. Lai, Z.B., Loh, W.K., Tamin, M.N.: Fatigue of Sn–4Ag–0.5Cu solder joints during cyclic bending of a BGA assembly. In: Proceedings of 11th International Conference on Electronics Materials and Packaging (EMAP) (2009)
8. Shaffiar, N.M., Loh, W.K., Kamsah, N., Tamin, M.N.: Continuum damage evolution in Pb-free solder joint under shear fatigue loadings. In: Proceedings of 12th Electronics Packaging Technology Conference (EPTC). pp. 304–308 (2010)

Closure

The ever-decreasing physical size of electronic packages and assemblies with BGA solder joints requires new and innovative methodologies in assessing and establishing their reliability. This book introduces a hybrid experimental-computational approach that combines experimental measurements and Finite Element (FE) simulations. Controlled experiments on solder materials and reflowed solder joint specimens generate measured data for the determination of properties, the extraction of material model parameter values and validation of the predictive models. FE simulations calculate internal states of deformation, strain, stress and damage in BGA packages with respect to solder joint reliability.

The minute size of BGA solder joints in electronic assemblies poses a great challenge in quantifying the mechanical behavior of the material and predicting performance. In this respect, FE simulation offers the following major advantages:

- Intrinsic properties of the bulk solder joint and solder/IMC interface can be determined using the hybrid experimental-computational approach. In this approach, the FE model is validated using measured response of the test system. Results of the validated model are then interpreted to establish useful physical properties.
- Given accurate constitutive and damage models for solder joints, FE simulation serves as an ultimate thinking tool in quantifying the mechanics and establishing the reliability of the solder joint.
- Validated FE simulation for the simplified mathematical model of a physical problem is ideal for the parametric study of BGA assemblies to quantify the sensitivity of selected materials and design variables to solder joint reliability.

Numerical simulation frameworks incorporating FE simulation as a primary predictive tool has reached an advanced level, both in terms of constitutive and damage model formulations and the efficiency of numerical codes. While the framework has been demonstrated in this book for solder joint reliability in BGA assemblies, it could be extended to other assemblies including those with leaded packages, copper pillar interconnections and through-silicon via (TSV).

M. N. Tamin and N. M. Shaffiar, *Solder Joint Reliability Assessment*,
Advanced Structured Materials 37, DOI: 10.1007/978-3-319-00092-3,

However, deployment of this approach as a design tool is hindered by a lack of measured data from controlled experiments and real applications. The much-needed information is on temperature- and the strain rate-dependent response of the solder/IMC interface, moisture-induced degradation of properties, grain coarsening of the bulk solder joint during shelf life, electro- and thermo-migration in the bulk solder joint and synergies among these phenomena in accelerating solder joint failure.

Since the FE model for specific problems is developed for solving the simplified mathematical model of the complex physical problem, the extrapolation of the FE-predicted results to situations beyond the assumed conditions should be exercised using knowledge of mechanics and the mechanisms of the system being examined. Common simplifications include the assumptions of *rigid* tools with respect to other materials being modeled, frictionless contact surfaces, and homogeneous IMC layers with uniform thickness and smooth morphology of the solder/IMC interface. Such assumption however is appropriate for the reliability model of solder joints at the meso-scale level.

The ability to correctly assess and accurately predict the level of material integrity provides precision in forecasting a material's behavior for the safe and reliable operation of the components and structures in harsh environments. In this respect, the hybrid experimental-computational methodology, as demonstrated throughout this book for assessment of solder joint reliability, could be extended with relative ease to newly-designed microelectronic assemblies and systems should relevant measured data be made available.

MIX
Papier aus verantwortungsvollen Quellen
Paper from responsible sources
FSC® C105338

If you have any concerns about our products,
you can contact us on
ProductSafety@springernature.com

In case Publisher is established outside the EU,
the EU authorized representative is:
Springer Nature Customer Service Center GmbH
Europaplatz 3, 69115 Heidelberg, Germany

Printed by Libri Plureos GmbH
in Hamburg, Germany